THE BOOK OF MAGIC SQUARES

JAIN

VOLUME 3

2nd EDITION
YEAR 2000

THE BOOK OF MAGIC SQUARES VOL 3 YEAR 2000 JAIN

- This unique and original "Draw-Your-Own-Mandala-Join-The-Dots-Colouring-In-Activity-Book is equally suitable for children and adults. Just by drawing a long, continuous line from 1→2→3 etc to the last No. of a Magic Sq., superb patterns appear, creating sacred symbols from many cultures and verging on Atomic Art.
- The traditional meaning of a M.Sq. is that the sums of its columns, rows and diagonals are Equal.
- Volume 1 (pub. 1990) covered M. Sqs. of 3, 4, 5, 6 and 7.
- Volume 2 (pub. 1996) covered M. Sqs. of 8, 9 and 10.
- Volume 3 (pub. 2000) covers M. Sqs. of 11, 12 and 16, in detail, giving many varied and creative ArtForms depicting Order amongst Chaos.

Beauty in Mathematics

- At the end of this Volume 3 there is a chapter giving all the graphical Solutions, as do Volumes 1 and 2, but unlike the others, this Volume has a special and in-depth Glossary of all Sacred Geometrical terms used.
- Also, unique to this Volume, is a chapter on "The Visual Multiplication Table and its Reduction Grid" (also known as The Vedic Square), that uses digital sums to create Atomic ArtForms, like the nuclear geometry of Platinum Crystal (shown here) for the first time in print, linking 20 years of Magic Square research to the lost Vedic Mathematical Sutra that Jain calls "... in the Next Dimension".
- Explore, extract what you can from this quarry of information and have Fun.

THE BOOK OF MAGIC SQUARES

JAIN
VOLUME 3

2nd EDITION
YEAR 2000

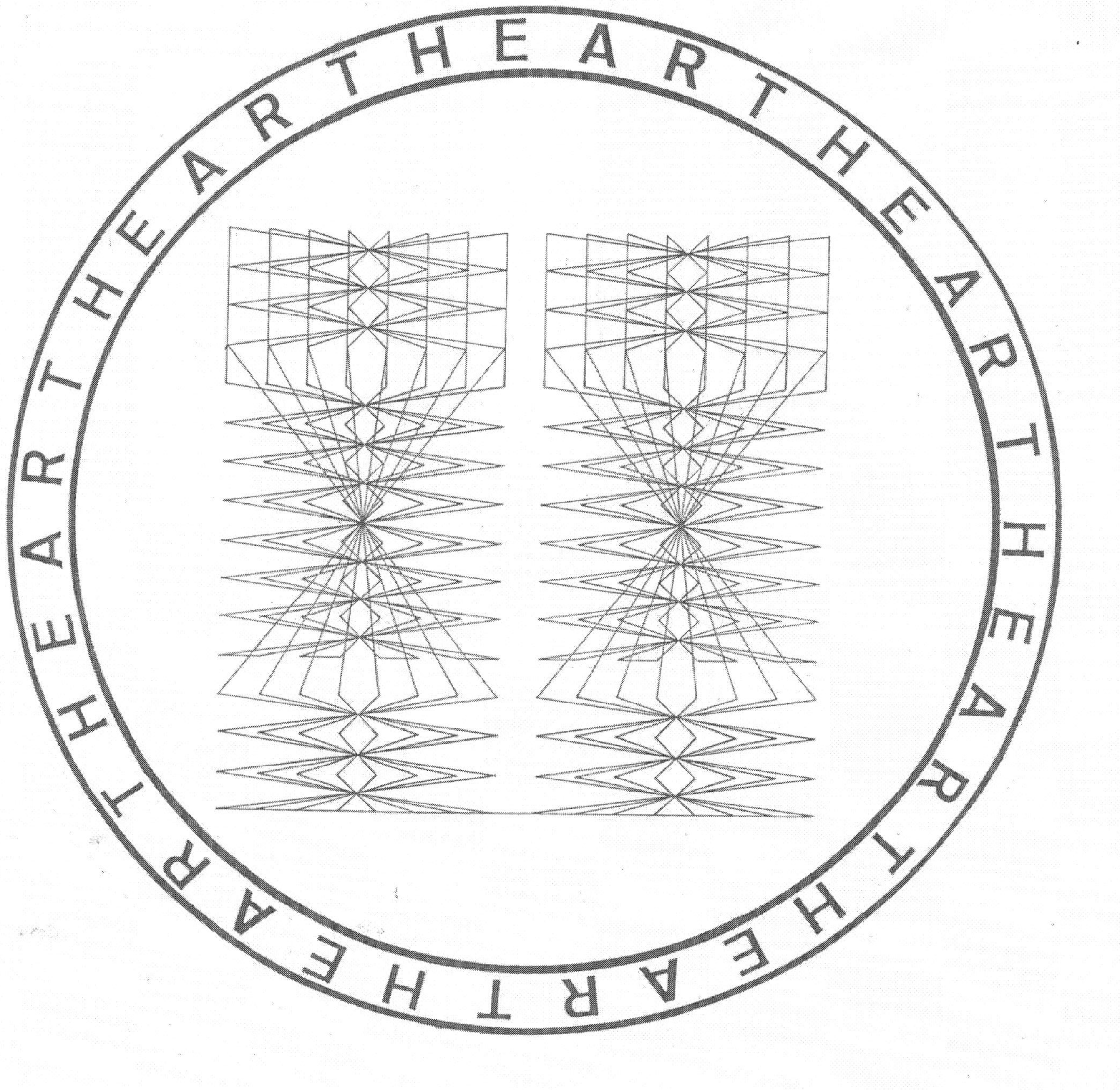

THE BOOK OF MAGIC SQUARES
VOLUME 3

JAIN ©

ISBN: 0 - 9594180 - 5 - 9

1st Edition (1996) 2nd Edition (2000)

Self Published by F.R.E.E.D.O.M.S. an acronym for:

For **R**esearch **E**xpressing **E**ssential **D**ata **O**f **M**agic **S**quares

For Correspondence, contact:
JAIN
c/o MAGIC SQUARE STUDIO
777 LEFT BANK RD, MULLUMBIMBY CREEK, Northern N.S.W. Australia, 2482.
Phone: 61 2 66 844409

email: jain@jainmathemagics.com
website: www.jainmathemagics.com

DEDICATED TO: ISHARNI JAINASUN AND HIS MOTHER LEIGH

né 25-2-2000 MULLUMBIMBY

God-dess-ence
I AM.

ISBN
0 – 9594180 – 5 – 9

To Live Life
To Love Life
To Live Love.

MAGIC SQUARE OFFICE

JAIN

Jain
Magic Square Studio
777 Left Bank Rd
Mullumbimby Creek
2482

Phone/Fax:
61 2 66844409

E-mail:
Jainmagic42@hotmail.com

THE BOOK OF MAGIC SQUARES

JAIN 2ND EDITION YEAR 2000 Volume 3

ISBN : 0 – 9594180 – 5 – 9

"We rise
by kneeling
We conquer
by surrendering
We gain
by giving up".

ALICE BAILEY

This Book
Belongs to :

CONTENTS Volume 3

Contents continued:

EDITORIAL

The Process Is The Goal.

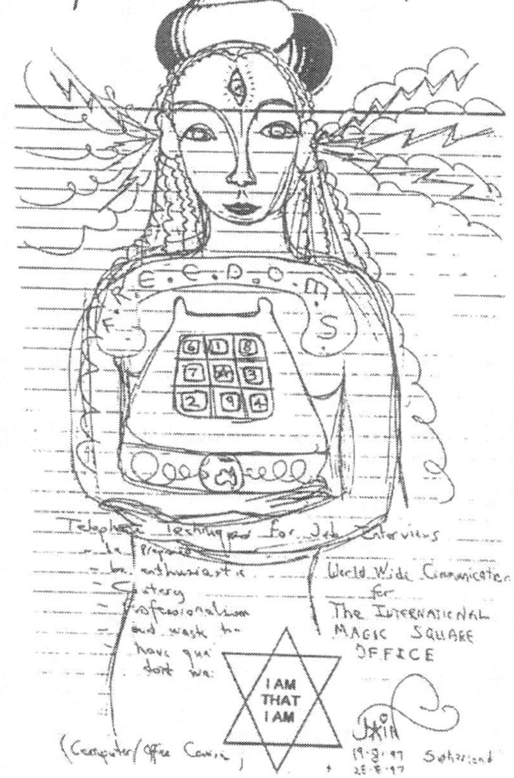

"Success is a Journey not a destination".
B. Sweetland.

CHANGES THAT HAVE BEEN MADE:

This book is half of a larger body of work that was titled: **THE BOOK OF MAGIC SQUARES . Vol 2. 1996.** The first edition was a 290 page compendium. Many readers have been overwhelmed by such a large book theory, graphics and activities on Magic Squareology. It is for this reason that it now exists as two smaller books: Volumes 2 and 3. This book, Vol 3, is the 2nd half of the original, and deals only with the Magic Squares of 11, 12, 16 and The Visual Multiplication Table. It is a sequel to My first book: **THE BOOK OF HARMONY SQUARES. Volume 1. 1990.** This latter title has also been changed to: **THE BOOK OF MAGIC SQUARES. Volume 1.** The contents are the same, only the title is difrnt.

is difrnt. The reason for this change, from "Harmony Sqs." to "Magic Sqs." is that 20 yrs ago some Christian Schools objected to the word "Magic" when I was teaching there. I feel now that I choose not to change or interfere with ancient titles, and I am proud to introduce such a body of work as "Magic Squares", despite the ongoing controversy. Currently the State of Queensland (Department of Education run by Fundamentalist Christians) has formally banned all my works, presence + books in their State, objecting to this word + its superstitious associations. Fortunately or unfortunately, Magic Square Artforms produce incredible atomic patterns formed from combinations of Christian Crosses + Eastern Swaztikas! This was revealed in Vol 1, page 110. Besides this, Magic Squares are enjoyed by all other State schools in Australia. If you are a Maths Teacher in Qld you can support me by employing me as an Authorized School Performer, my unique theatrical show being known as "MATHEMAGICS".

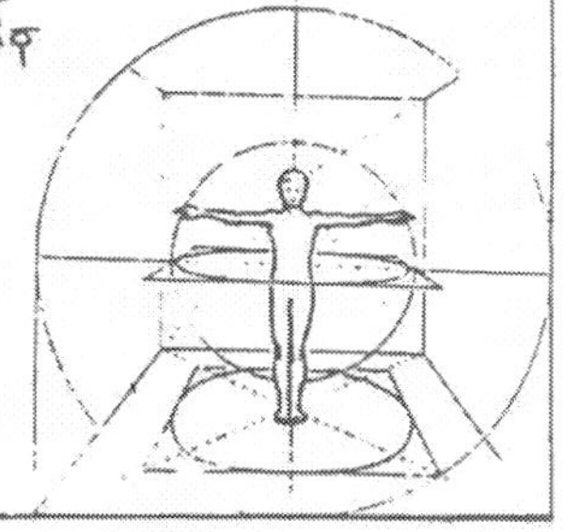

This Compendium contains :

- ◇ **174** pages.
- ◇ **94** original drawings.
- ◇ **37** geometric / M. Sq. Activities to draw.

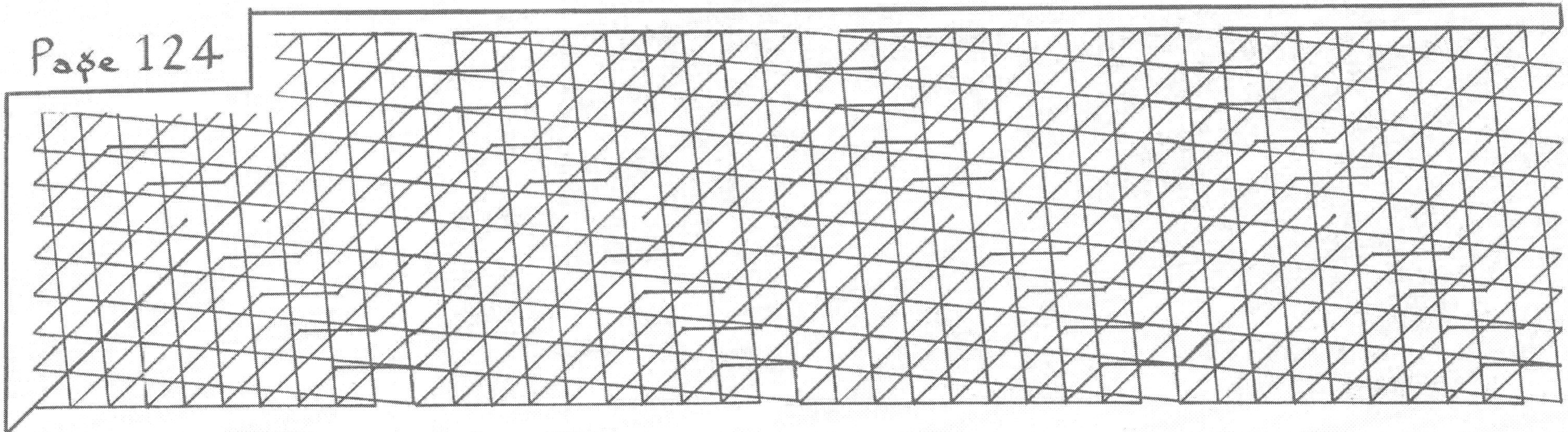

Chapter ELEVEN EXPLORING OTHER DIMENSIONS THE 11th DIMENSION

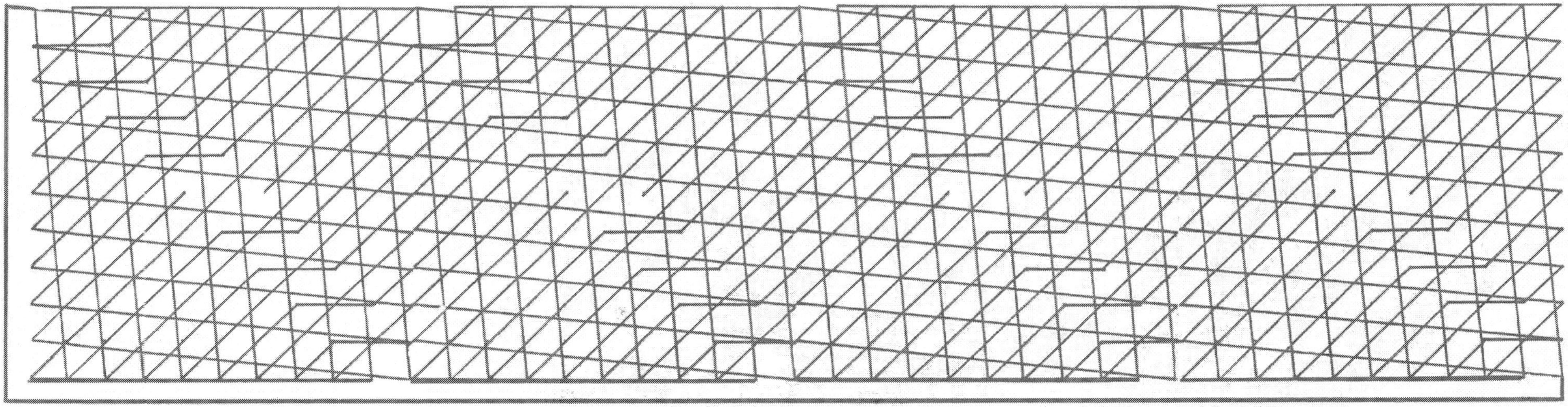

THE ESSENE'S MAGIC SQUARE OF ELEVEN

The Essenes were a X^t^ian brotherhood-sisterhood into which Iesus the Christ (or Jeshua ben Josef) was born into. The Essenes then were also known as The Nazarites, Nasirs or Yessenes. They practised community of goods, vegetarianism & fasting in their desert lands. They wrote of Iesus' teachings on scrolls of skins known as "The Gospel of The Holy Twelve" which sat in an urn in Tibet for nearly 1900 years, recently translated from the Aramaic, it contains Majik Squares of 3, 7 & 11 which were used as lengthy Parables to illustrate the Divine Kingdom / Hierachy / Centre / Unity.

"The MAJIK SQ. OF ELEVEN PERFECTED"
(ECCLESIAE MILITANTIS SIGILLUM)

The M. Sq. of 11 (Fig 110) has been found wonderfully to explicate (i.e make clear) symbolically at least, to the mystical, the meaning of the Divine Order. The form exactly illustrates what the Lord in symbol taught to his disciples of the bringing forth of order out of disorder, perfection out of imperfection, & out of deficiency fullness. This square was known to the early gnostics who revered its central crosses summing the same as the columns etc.

56	117	46	107	36	97	26	87	16	77	6
7	57	118	47	108	37	98	27	88	17	67
68	8	58	119	48	109	38	99	28	78	18
19	69	9	59	120	49	110	39	89	29	79
80	20	70	10	60	121	50	100	40	90	30
31	81	21	71	11	61	111	51	101	41	91
92	32	82	22	72	1	62	112	52	102	42
43	93	33	83	12	73	2	63	113	53	103
104	44	94	23	84	13	74	3	64	114	54
55	105	34	95	24	85	14	75	4	65	115
116	45	106	35	96	25	86	15	76	5	66

FIG 110

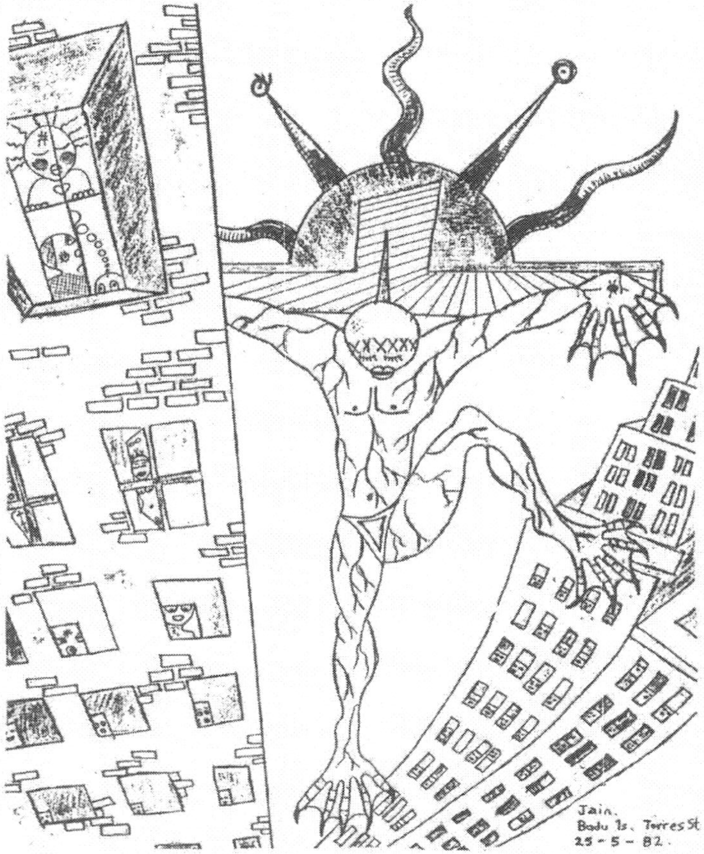

THE GOSPEL OF THE HOLY TWELVE.

Fig 110 on the previous page has 33 different Summations of 671.

The sum of all the terms (i.e Sigma "Σ" = 1 + 2 + 3 + ... + 120 + 121) = 7,381.

There are, of significance, 4 central crosses that also have the same Constant (C = 671) e.g.: 85 + 13 + 73 + 1 + 61 + 121 + 49 + 109 + 37 + 11 + 111 ;
13 + 73 + 1 + 61 + 121 + 49 + 109 + 111 + 51 + 11 + 71 ;
73 + 1 + 61 + 121 + 49 + 111 + 51 + 101 + 11 + 71 + 21 &
1 + 61 + 121 + 111 + 51 + 101 + 41 + 11 + 71 + 21 + 81.

On the following page is The Majik Square of Eleven Parable from "The Gospel of the Holy Twelve". For more information refer to my first majik square mandaloid book : "The Book of Harmony Squares" Volume One, in Chapter 7 which gives a detailed account of "The Parable of The 7 Palms" all about Jesus' story of The Venusian Majik Square of Seven.

.....17 And the Lord
gathered together all his
disciples in a certain place. And he
said unto them, Can ye make perfect-
ion to appear out of that which is imper-
fect ? Can ye bring order out of disorder
? And they said, Lord, we cannot.
18. And he placed them according to the
number of each in a four-square order, each
side lacking one of twelve, and this he did,
knowing who should betray him (who should be
counted one of them, but was not of them).
19. The first in the seventh rank from above
in the middle, and the last in the seventh
from below, and him that was neither first
nor last did he make the Centre of all, and the
rest according to a Divine order did he
place them, each finding his own place, so
those which were above, were even
as those which were below, and
those which were below as those
which were above, and the
left side was equal to
the right side, and
the right side to
the left, according to the
sum of their numbers
20. And he said, See you how
ye stand ? I say unto you, In
like manner is the order of the king-
dom, and the One who ruleth all is in
your midst, and he is the centre, and
with him are the hundred and twenty, the
elect of Israel, and after them cometh the
hundred and forty and four thousand, the
elect of the Gentiles, who are their brethren.

(This is an extract from :
"THE GOSPEL OF THE HOLY TWELVE
Lection LXVIII
A MAJIK SQUARE PARABLE by
IESUS, the Essene)".

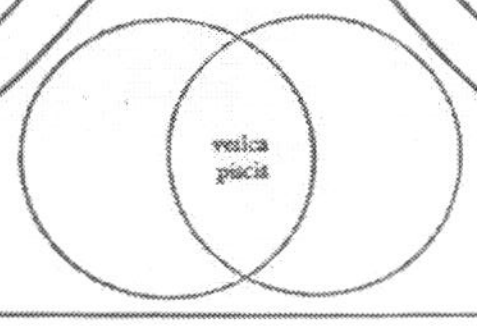

I went to my Lebanon during a time of war and there I wr-ote of Pe-ace et er n a l.

In the sacred space of Fig 111 carefully & meditatively construct the Essene Harmony Square of Eleven at 0°. Hold clear in the throne of your mind this clichéd affirm-ation : "The Pen is mightier than the Sword".

Fig 111

Let Fig 111a (shown in "The Solutions") refer to the method of Transposition for how to construct fig 111

In the space of Fig 112 construct the ESSENE MAJIK SQUARE YANTRAM at 0° + 90°. (That means re-writing Fig 110 on a separate piece of paper to allow the 90° rotation).

To avoid flicking back to Fig 110 to follow the sequential order of the numbers, just rewrite it again.

Page 131

Fig 112

The Opening of the Doorway

11:11 is a pre-encoded trigger placed within our cellular memory banks prior to our descent into matter which, when activated, signifies that our time of completion is near.

The two yantra shown are respectively: THE ESSENE MAJIK SQUARE OF ELEVEN at 4 x 45° (i.e: Fig 113) and 16 x 11·25° (i.e: Fig 114). Both are drawn by hand & need to be computerized for futher precision. You may have that task.

The date for the next planetary activation is happening. NOW !!

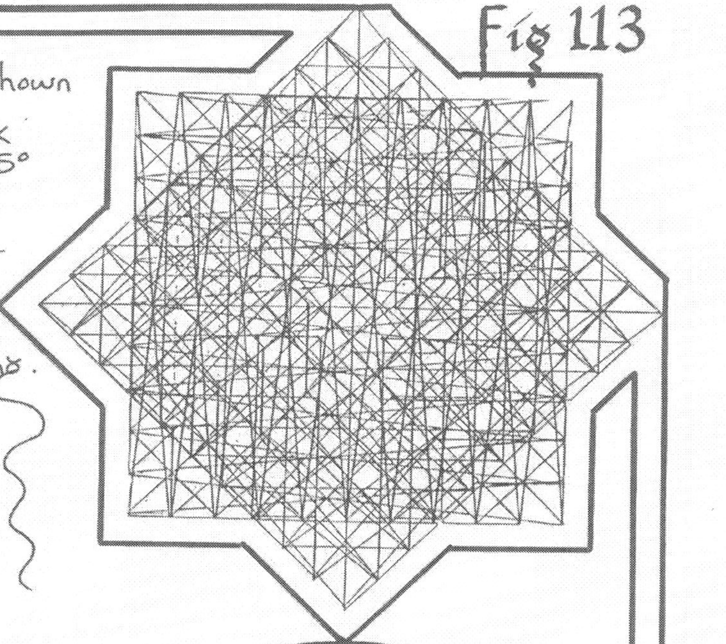

Fig 113

Fig 114

The 11:11 is hereby being activated...

Showing the relationship of a grid polar square to the geographic pole. Each grid has a similar pattern.
The pole of each grid is set at a different latitude and longitude.

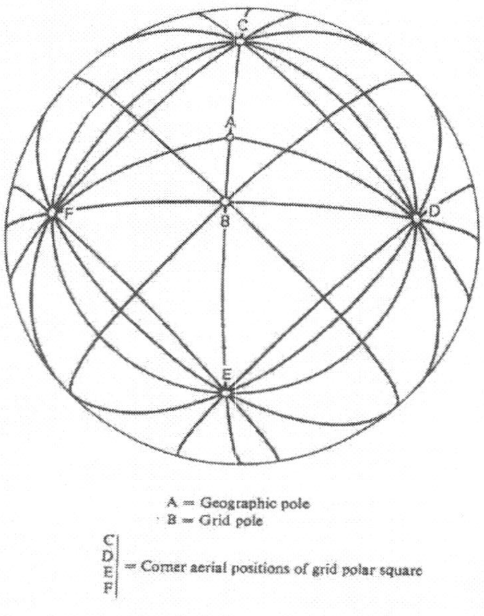

Showing the relationship of grid polar squares A, B, and C. The polar squares are orientated in reciprocal positions around both the north and south geographic poles.

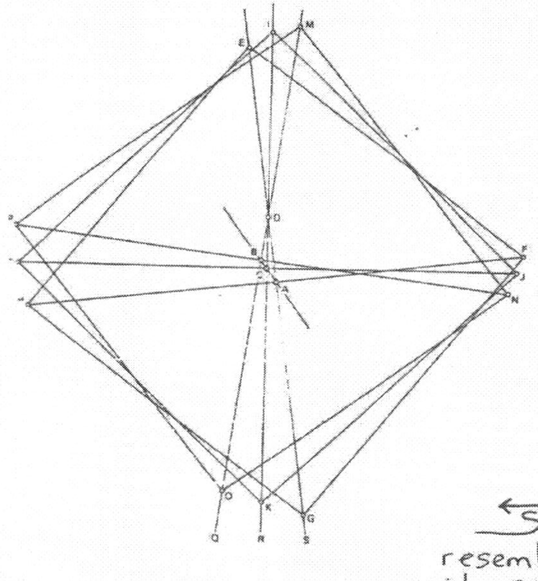

Fig 115

Showing the harmonic wave-form which creates an atom & anti-matter, in alternate pulses.

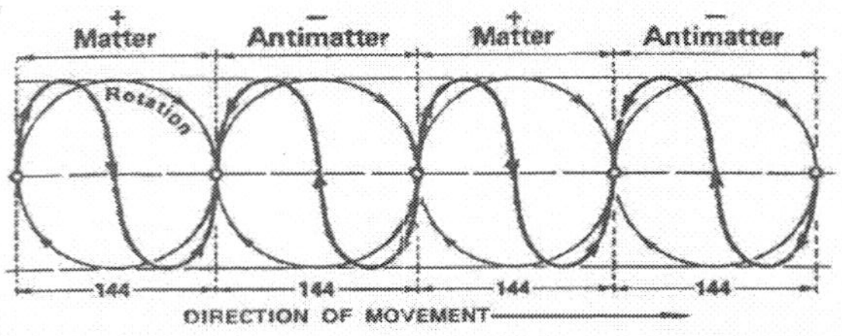

Captain Bruce Cathie

" I began my research in 1965. I had observed several unidentified flying objects while carrying out my flying duties with NAC (the New Zealand national airline), & eventually discovered that their movements conformed to a worldwide grid pattern. At first I believed the grid structure to be artificially produced, but now I know it to be the natural form through which all physical matter is manifested within the universe.

Over the years I have been able to produce unified equations which demonstrate the harmonic affinity between the angular velocity of light & all the physical substances."

Notice how the Essene Majik Square of Eleven at 0° Yantram resembles this diagram of B. Cathie's grid polar squares: (Fig 115).

Does it suggest to you that this yantram that you drew in Fig 111 relates to the Earth's auric electro-magnetic grid ley lines that is the basis of the Universal Free Energy System? Tesla proved that by placing certain "antennae" in harmony with this invisible grid, we, as a planet, could tap into "Eternal Free Energy". He taught that when the so called "electric light bulb" was discovered, it was meant to be an "eternal light bulb", but the tyranny of the secret government has made them short-lived for more dollars$dollars$dollars$dollars$dollars$dollars. dollars$dollars$dollars$dollars$dollars$dollars.

AN ALTERNATIVE PATTERN HARMONY SQUARE OF ELEVEN

68	80	92	104	116	7	19	31	43	55	56
81	93	105	117	8	20	32	44	45	57	69
94	106	118	9	21	33	34	46	58	70	82
107	119	10	22	23	35	47	59	71	83	95
120	11	12	24	36	48	60	72	84	96	108
1	13	25	37	49	61	73	85	97	109	121
14	26	38	50	62	74	86	98	110	111	2
27	39	51	63	75	87	99	100	112	3	15
40	52	64	76	88	89	101	113	4	16	28
53	65	77	78	90	102	114	5	17	29	41
66	67	79	91	103	115	6	18	30	42	54

Fig 116

Introduction

There is no other redemption for mankind than to realize their divinity. You are the seeds of this realization.

I AM RAMTHA, A SOVEREIGN ENTITY who lived a long time ago upon this plane called Earth, or Terra. In that life I did not die; I ascended, for I learned to harness the power of my mind and to take my body with me into an unseen dimension of life. In doing so, I realized an existence of unlimited freedom, unlimited joy, unlimited life. Others who lived here after me have also ascended.

I am now part of an unseen brotherhood who love mankind, greatly. We are your brothers who hear your prayers and your meditations and observe your movements to and fro. We are those who lived here as man and experienced the despair, the sorrow, and the joy that all of you have known. Yet we learned to master and transcend the limitations of the human experience to realize a grander state of being.

I have come to tell you that you are very important and precious to us, because the life that flows through you and the thought that is coming to every one of you—however you entertain it—is the intelligence and life-force that you have termed God. It is this essence that connects all of us, not only to those upon your plane, but to those in untold universes which you have not yet the eyes to see.

I am here to remind you of a heritage which most of you forgot, long, long ago. I have come to give you a loftier perspective from which you may reason and understand that you are, indeed, divine and immortal entities who have always been loved and supported by the essence called God. I am here to help you realize that only you, through your sublime intelligence, have created every real-

Jain.

In the sacred space of Fig 117 construct the Alternative Pattern Majik Square of Eleven (Fig 116) Yantram at 0°.

Referring to Figs 49 & 50 you are required to construct a m. Sq. of 11 x 11 via the Method of Transposition. Upon a sheet of paper, beginning with a Natural Square, create Fig 110.

"Live not in the past tense. Live not in the future tense, but rather live in the pleasant tense".

Fig 117

Chapter Twelve.

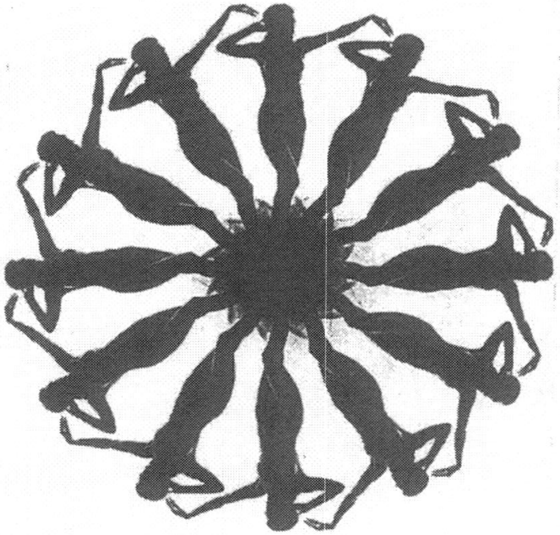

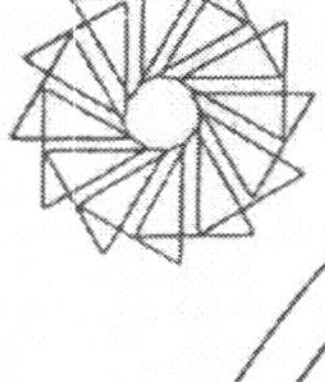

This chapter 12 will have a few M. Sqs. of 12 for you to draw & color-in, but there is also a visual emphasis for any image that has 12 rays, points or sides. This is a chapter on "TWELVENESS" e.g: Fig 118 shown here is a computerized yantram of 12 circles. This source of visual graphics is the format for this whole book, previous & succeeding volumes, & is an important means of stimulus into a pedantically special-ized mathematical topic such as Majik Sq. Theory.

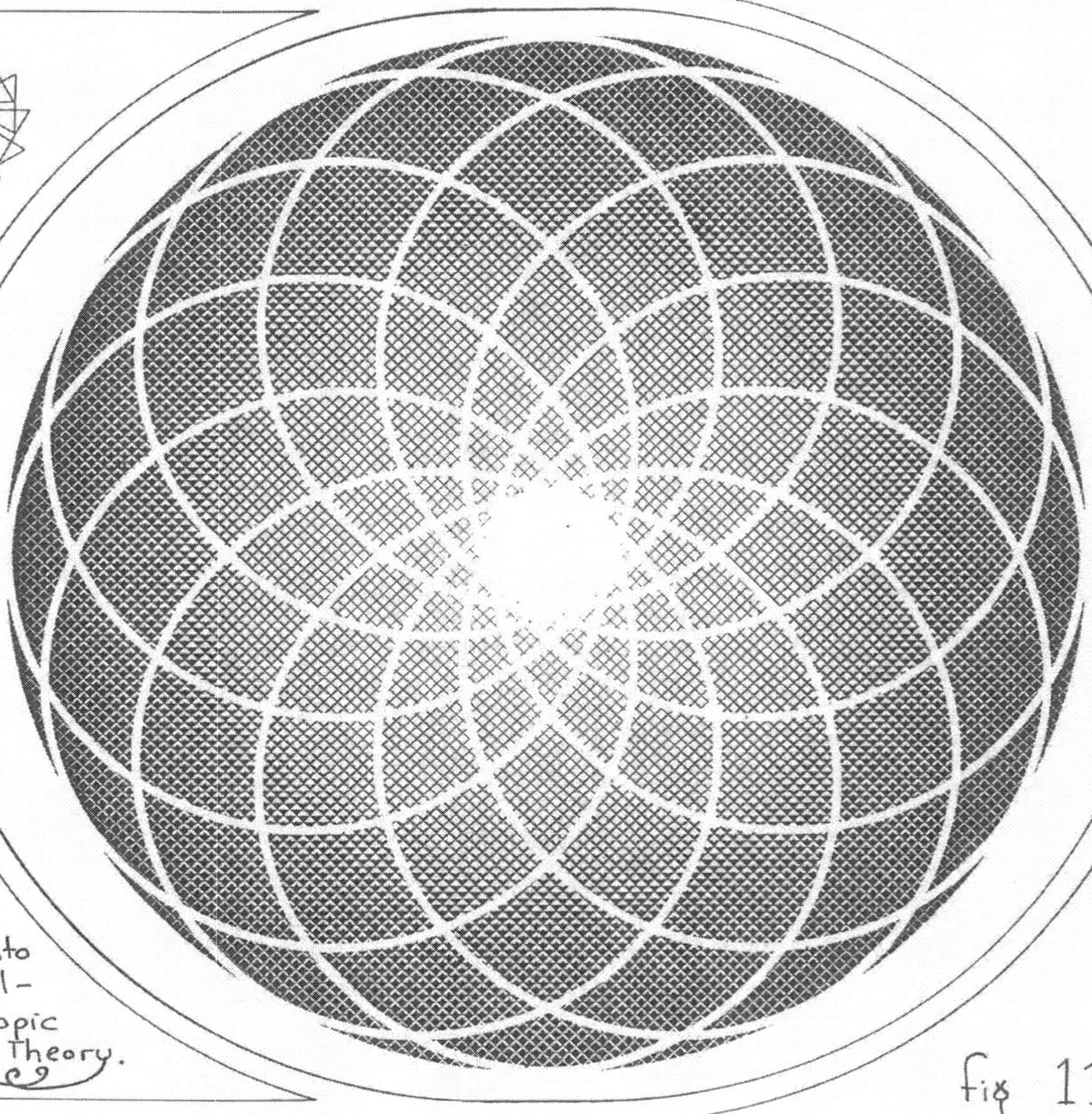

Fig 118

"The DIVINE QABAHLISTIC SIGNET STAR".

Fig 119

Fig 119 is a secret symbol of the Rosicrucians: The Star of Solomon (falsely known as The Star of David) where the sums of the 3 columns add up to 12 i.e: 1+4+7, 3+4+5, 2+4+6. Also, the complementary pairs or the Polar Opposites e.g 1 and 7 have a sum that is double the Centre No. Also each Triangle sums 12 eg 1+5+6.

With any cosmic array of numbers, the challenge is how to transform it or re-express its vibration in an artistic, mandalaic form. Here is a Circle → with 7 dots to begin the process. Cont...

Fig 120

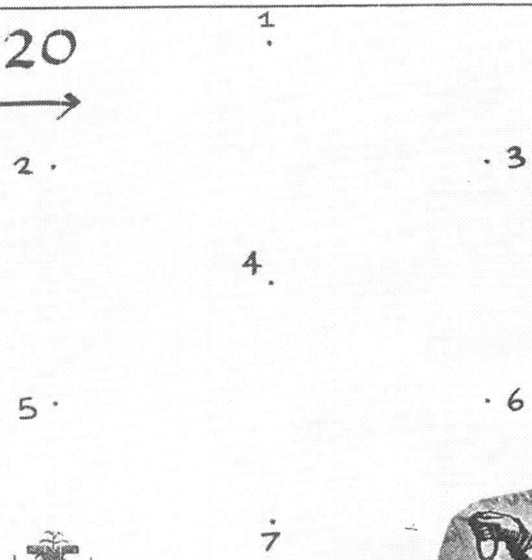

In the Space of Fig 120 join a line from 1 → 2 → 3 → 4 → 5 → 6 → 7 → 1. This is the arrangement of Nos. as shown in Fig 119.

In the Space of Fig 121 draw this 0° yantra again. Then you will superimpose two more rotations at 30°. To do this first 30° rotation join the Nos: 3 → 1 → 6 → 4 → 2 → 7 → 5 → 3. cont...

FIG 121

To superimpose the second 30° rotation join the Nos. : 6 → 3 → 7 → 4 → 1 → 5 → 2 → 6.

In this book's jargon, this design in Fig 121 can be technically called "The Divine Qabahlistic Signet Star Yantram at 0° + 30° + 60°"

NEWS from the ART WORLD

12

Page for the Contemplation of Arabian Art

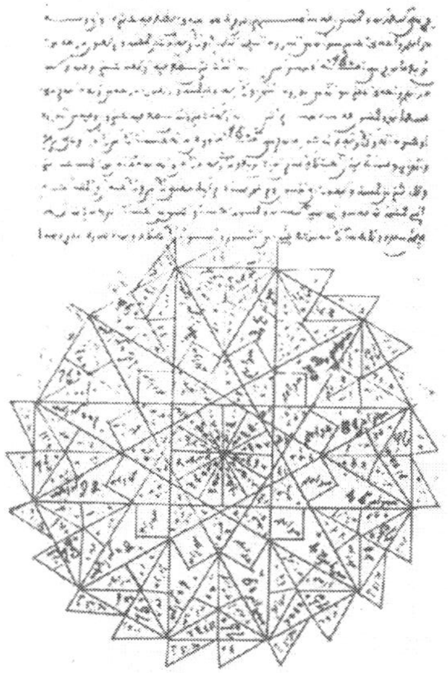

the Islamic vision of things, it becomes evident that number and figure exist on three levels of reality: in the Divine Intellect as archetypes in the principial domain; in the intermediate world of the mind, which Nichomachus refers to as 'scientific'; and in the external world, corresponding to 'concrete', quantitative numbers and figures. The modern world knows only the second and third levels, whereas Islam has remained always aware of all three. The use of mathematics in Islamic art and architecture is not only the result of an 'aniconic' tendency away from the 'concrete' toward the 'abstract'. Rather, it is also a means whereby the archetypes are reflected upon the material plane, making the material transparent and capable of acting as a ladder toward the spiritual realities, which are the most concrete realities of all, realities compared to which physical realities are no more than abstractions. The use of mathematics in Islamic art is a way whereby the material is sacralized by virtue of reflecting the archetypal world. It is also the means whereby man becomes aware of the origin as well as the fundamental structure of the physical world that surrounds him and is able to penetrate into the very mystery of God's creation.[33]

Mathematics in Islam has not been bound to the world of 'matter' as is modern mathematics. It has in fact been related more to the world of life forms and beyond that to the archetypal world. As a result it has been able to reveal the principle of the physical world rather than the structure of its constitutive parts as in modern physics. Moreover, it has been able to aid in the realization of a harmony, balance and awareness of the effusion of multiplicity from Unity and the return of all multiplicity to Unity which characterize Islamic spirituality and are manifested in the most direct manner in Islamic art and architecture. Nowhere is the sacred character of mathematics in the Islamic world view more evident than in art, where with the help of geometry and arithmetic matter is ennobled and a sacred ambience created wherein is directly reflected the ubiquitous Presence of the One in the many.

Page for the Contemplation of Celtic Art.

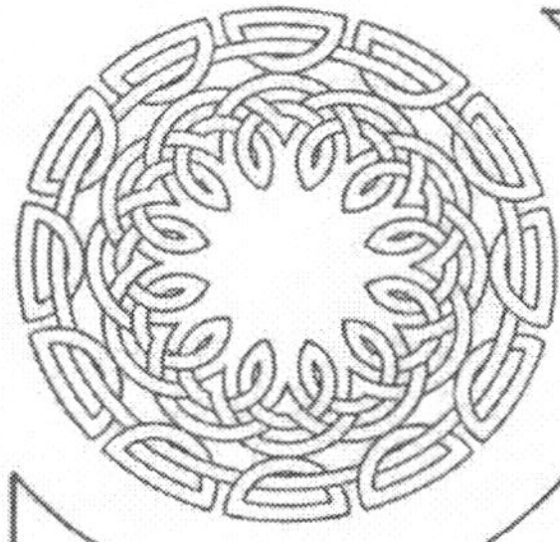

In the space of Fig 122 find the beginning of your pathway & walk the Maze.

Fig 122

Fig 123

1	143	142	4	5	139	138	8	9	135	134	12
132	14	15	129	128	18	19	125	124	22	23	121
120	26	27	117	116	30	31	113	112	34	35	109
37	107	106	40	41	103	102	44	45	99	98	48
49	95	94	52	53	91	90	56	57	87	86	60
84	62	63	81	80	66	67	77	76	70	71	73
72	74	75	69	68	78	79	65	64	82	83	61
85	59	58	88	89	55	54	92	93	51	50	96
97	47	46	100	101	43	42	104	105	39	38	108
36	110	111	33	32	114	115	29	28	118	119	25
24	122	123	21	20	126	127	17	16	130	131	13
133	11	10	136	137	7	6	140	141	3	2	144

Michelangelo's design of a continuous pathway in quadrangle of the Capitol, Rome, from an engraving by DuPérac in 1569.

Fig 123 has its columns, rows & diagonals summing 870.

In the space of Fig 124 carefully construct this Majik Square of Twelve's Yantra at 0°. The result is quite spectacular where many lines converge upon two central focci. For the desired effect, a very fine pencil or ink pen is required. As usual, refer to the Solutions in the back to check that your Yantra is identical.

Fig 124

FIG 125 →

This box of numbers is called a **COMPOSITE** Majik Square of 12 (fig 125). It is made up of 9 Majik Squares of 4 based on :

16	2	3	13
5	11	10	8
9	7	6	12
4	14	15	1

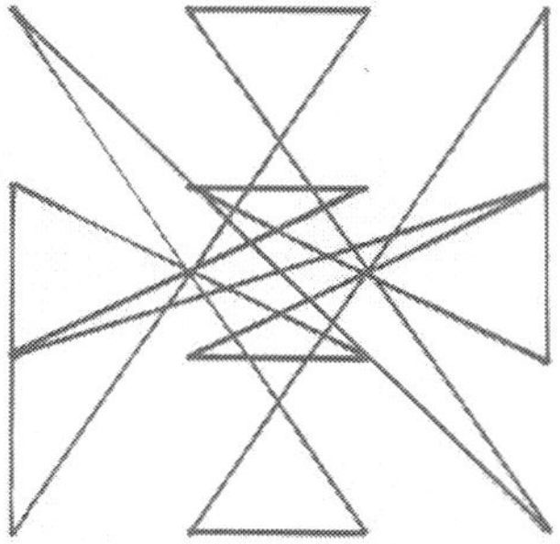

When the first 16 Nos. are drawn in the area of "1" of the Lo-Shu (Majik Square of 3) the movement is towards the next cell marked "2". 16 Nos. are drawn then move to "3" etc.

6	1	8
7	5	3
2	9	4

81	95	94	84	1	15	14	4	113	127	126	116
92	86	87	89	12	6	7	9	124	118	119	121
88	90	91	85	8	10	11	5	120	122	123	117
93	83	82	96	13	3	2	16	125	115	114	128
97	111	110	100	65	79	78	68	33	47	46	36
108	102	103	105	76	70	71	73	44	38	39	41
104	106	107	101	72	74	75	69	40	42	43	37
109	99	98	112	77	67	66	80	45	35	34	48
17	31	30	20	129	143	142	132	49	63	62	52
28	22	23	25	140	134	135	137	60	54	55	57
24	26	27	21	136	138	139	133	56	58	59	53
29	19	18	32	141	131	130	144	61	51	50	64

SECTION TWELVE

In the space of Fig 126 construct the Composite Majik Square of 12 Yantram at 0°.

Fig 126

FIG 127 →

84	65	68	47	82	93	70	45	80	95	98	43
67	76	83	92	69	46	81	94	71	44	51	96
64	85	66	77	48	35	128	79	50	97	42	99
111	30	75	34	91	78	49	72	129	52	133	16
86	63	110	31	74	127	36	53	132	17	100	41
29	112	33	90	59	54	73	104	37	130	15	134
62	87	58	109	32	1	126	131	18	105	40	101
113	28	89	60	125	108	55	2	103	38	135	14
88	61	124	57	144	121	6	19	106	3	102	39
27	114	25	122	7	56	107	120	5	138	13	136
24	123	116	143	22	9	118	141	20	11	4	139
115	26	23	8	117	142	21	10	119	140	137	12

Twelve by Twelve
KNIGHT'S TOUR
(SEMI-MAJIK SQ).

Beginning from cell No. 1 all moves are "L - shaped" and the last move (cell No. 144) returns back to the beginning. No cell is visited twice. The summations of all columns & rows are 870 except for the diagonals. By definition Fig 127 is not a true M.Sq. ∴ called "SEMI", but it has been included, for its yantram at 0° which you will now draw in the space of Fig 128 is especially delightful.

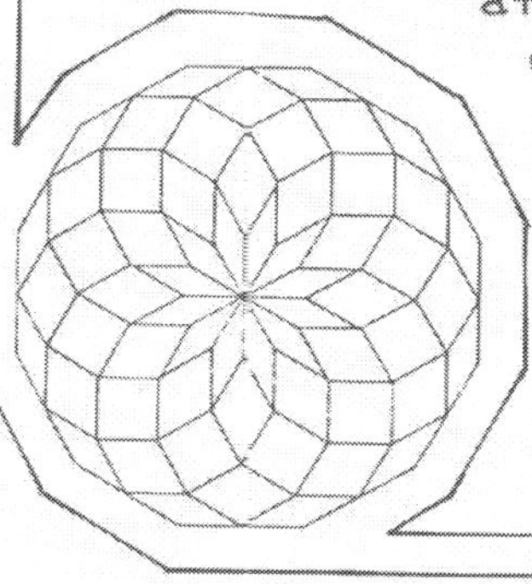

Page 147

FIG 128

from THE GOSPEL OF THE HOLY TWELVE;

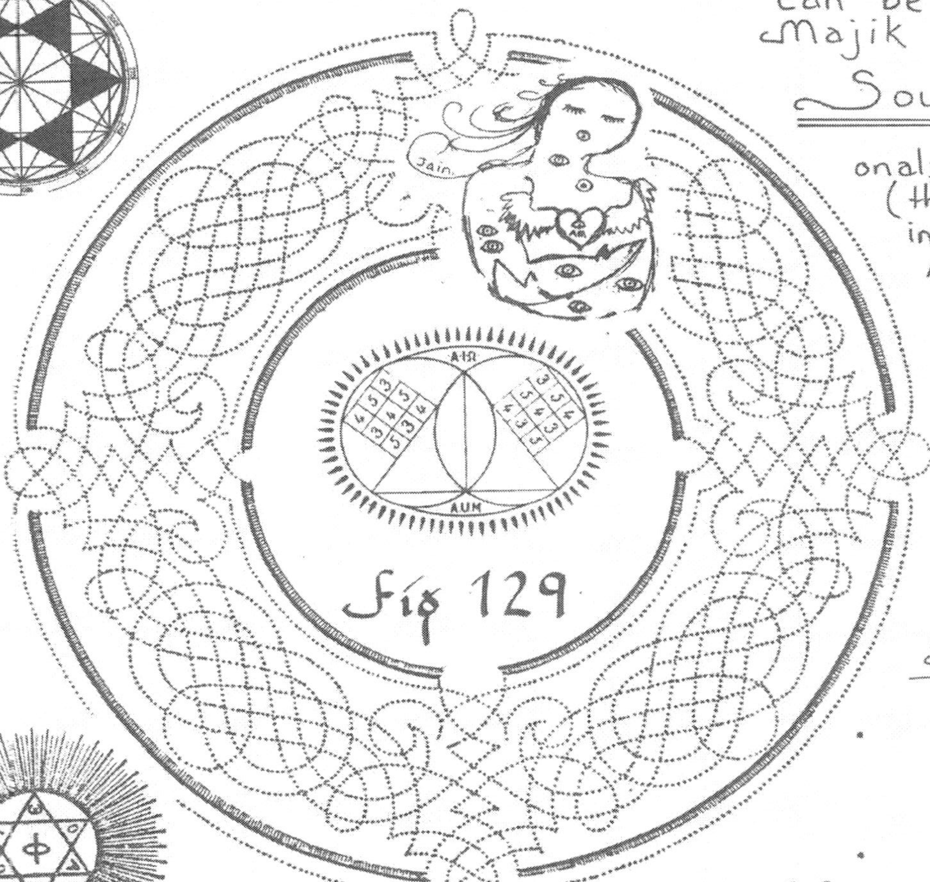

The STAR of SOLOMON.

can be arrived at from two different Majik Square Sources :

3	5	4
5	4	3
4	3	5

Source 1 : All Columns, rows & diagonals have a Sum of 12, (that is why this M. Sq. is included in this Chapter 12).

All opposing pairs are double the Centre No.

The Squares shown in Fig 129 are a 180° reflection of each other. Their digits "3", "4", "5" relate to the Pythagorean Right-Angled Theorem $3^2 + 4^2 = 5^2$ as shown in Fig 129.

Let us study the M. Sq. on the right hand side of Fig 129.

In the space of Fig 130 below connect the 3 points of "3" (a triangle).

Superimpose the 3 points of "4" & then the 3 points of "5" (a triangle).

Then, upon these 9 dots, do it at 0° + 90°.

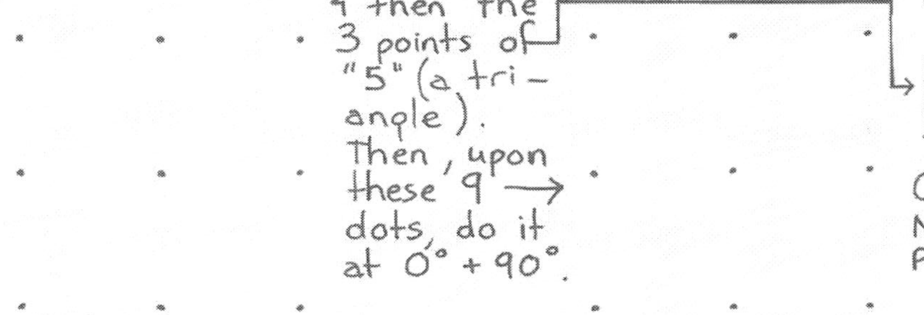

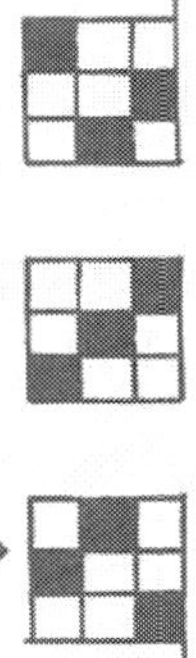

Cont. Next Page.

Fig 130

Cont... "The STAR of SOLOMON YANTRAM"

OR "The Pentacle of King Rabbi Solomon".

Source 2 :

6	1	8
7	5	3
2	9	4

Lo - Shu or m. Sq. of 3 ←

The Natural Square of 3 →

1	2	3
4	5	6
7	8	9

MAKING TALISMANS

THE PENTACLE OF RABBI SOLOMON THE KING

Fig 134

Fig 131 →

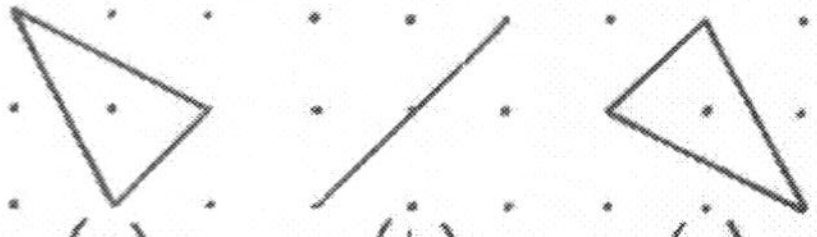

(a) (b) (c)

Observe the 3 squares on your left (Figs 131 a, b, c). Look at the top line of the m. Sq. of 3 which reads: "6 1 8" then find where these 3 numbers are positioned in the Natural Square of 3. The position of "618" in the Natural Square is shown as Fig 131a

Fig 131b shows the row of "7 5 3" &

Fig 131c shows the row of "2 9 4".

In the space of Fig 132 superimpose the ← 3 diagrams of Fig 131 a+b+c to form an <u>elongated Star of Solomon</u> (n.b: both triΔngles are approximately in "Golden phi 1:1·618 Proportions). Then in the space of Fig 133 → draw this Rabbi's Golden Star at 0°+90°. NB: This Yantram is very similar to the ancient diagram show above in Fig 134.

Fig 132

Fig 133

Surface of Silicon Crystal

Fig 133a Created by the Majik Sq. of Three.

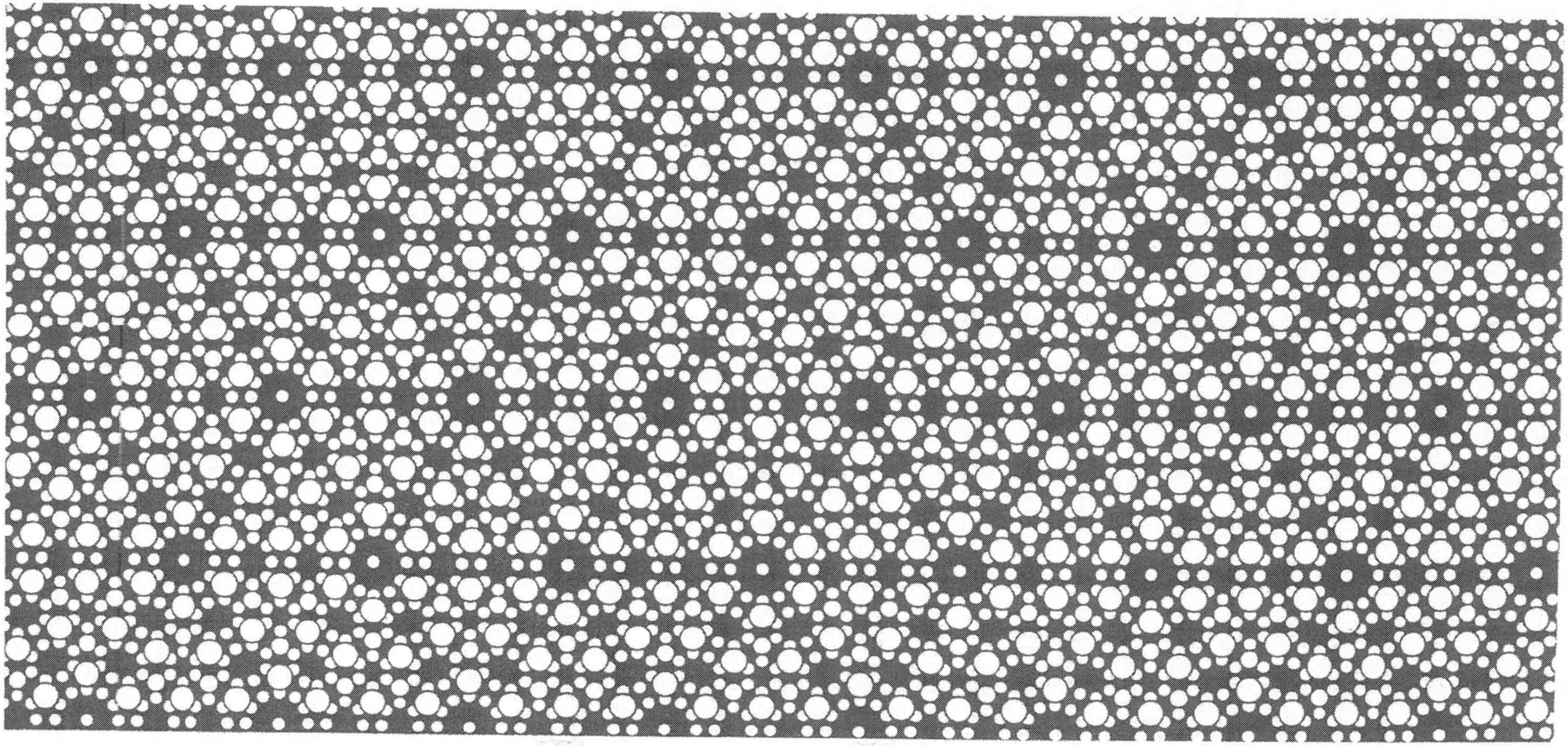

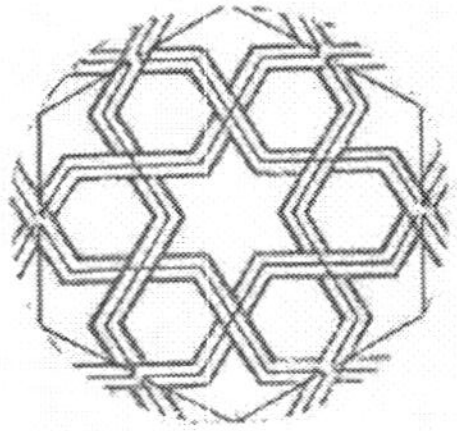

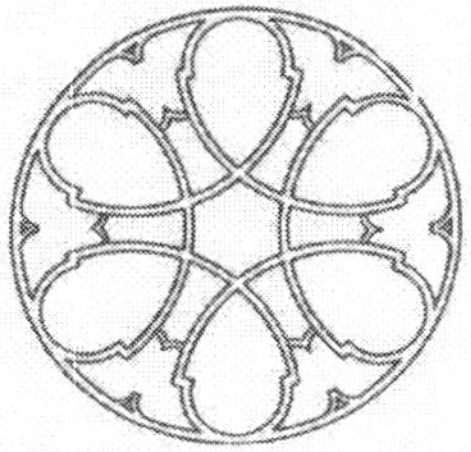

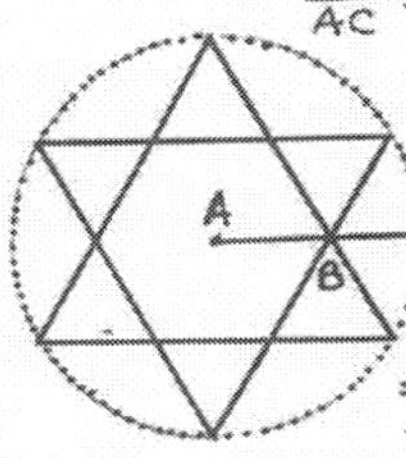

Fig 133a shows the surface of Silicon Crystal magnified approximately 100,000,000 times. (from a hand drawn poster by David Mankey, 1991). Relate this to figs 119, 129, 133 & 134 & 135. "... only since the recent invention of the Scanning Tunnelling Microscope, linked to modern computer development, that objective views of the mysterious lattice networks of energy known as electron fields have come before human vision. Small circles represent electron fields which are arranged tetrahedrally around their nuclei & bond together basic units of the crystal. This primordial pattern is the action of sub-atomic wave/particles at the quantum level of existence, and so is at the limit of possible microscopic view of whole particles. Most of Earth's surface crust is Silicon. It is the most abundant element in our immediate outer world". (D. Mankey).

Thus the Seal of Solomon (Fig 134) created by the M.Sq. of 3 analysed row by row (Fig 133) has always secretly represented the power of the atoms, the power of Crystals, the power of Spiritual Geometry. But remember, that which is important, is rather THE POWER OF LOVE, NOT THE LOVE OF POWER.

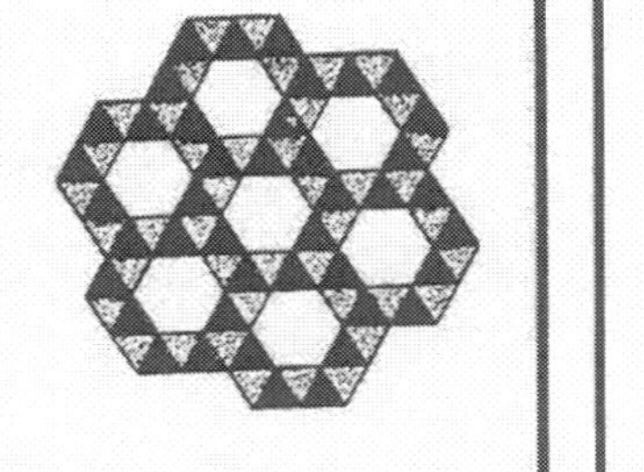

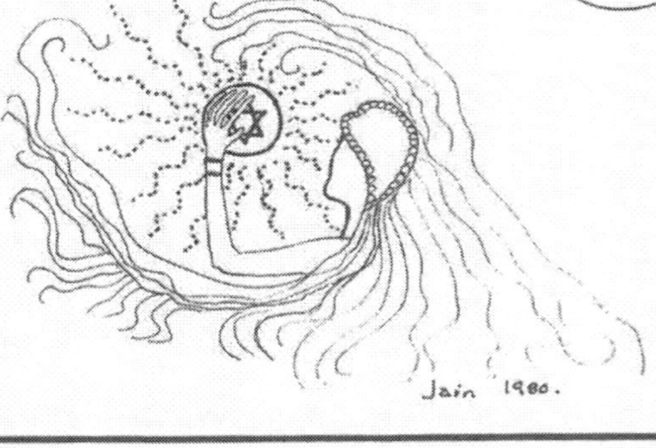

fig 132 a

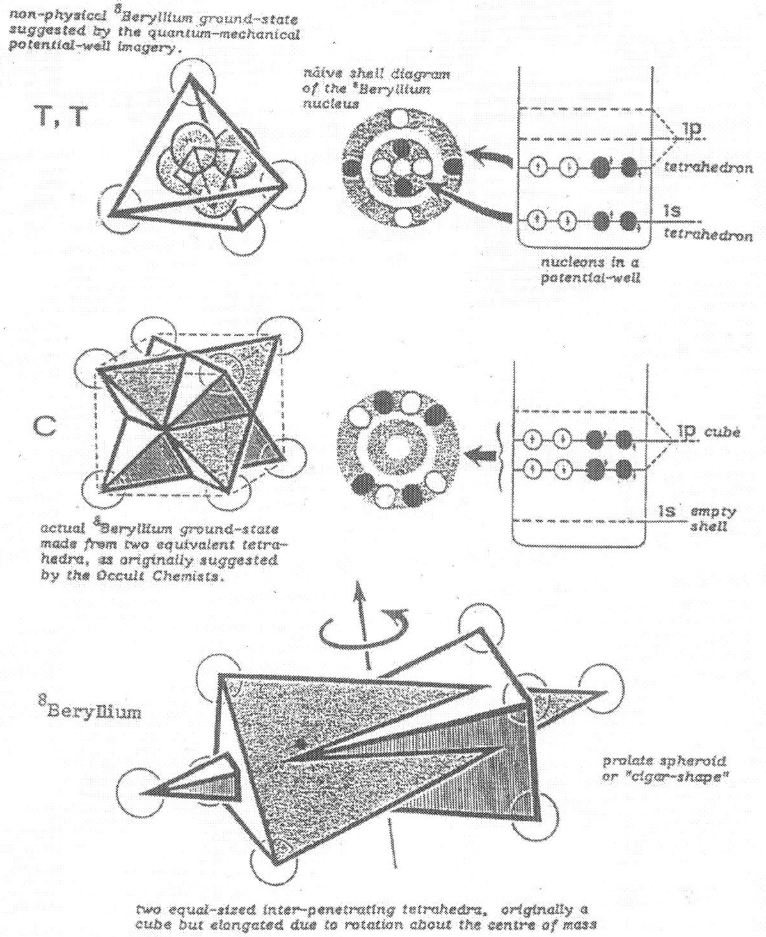

A Major Realization !

The final diagram shows two inter-penetrating tetrahedra: (atomic structure of Beryllium) ! This diagram is identical to the three-dimensional view of fig 132 whose Source is the centre of the Tibetan Calendar. That is why this knowledge has been kept secret for thousands of years, that the Magik Square of Three (fig 132) when viewed holographically, depicts clearly the atomic structure of Beryllium. Not only that, the major realization is that this hologram is the underlying shape of MERKABAH VEHICLES which will be explained.

The diagrams on this page are taken from Chris Illert's "Alchemy Today" Vol 2. The top diagram shows the error of quantum-mechanics. The central diagram shows the true shape, Notice how the interlaced tetrahedra form a cube when the 8 vertices are joined. It is essential for you to study the PLATONIC SOLIDS which have been carved in stone 20,000 years ago. So they don't really belong to Plato circa 300 BC. Now what happens when these two equal-sized inter-penetrating tetrahedra are rotated about their centre of mass ? They elongate to the golden mean proportions ! Notice that even this Line of Rotation is shown in Fig 132. This is the principle of the Merkabah, the spinning of the sacred Geometries at 90° Phase Shift of Consciousness. So what is the MERKABAH ?

It is the Divine Light Vehicle used by the Masters to probe & reach the faithful in the many dimensions of the Divine Mind. The MerKaBah can take on many forms of a brilliant briolette (pear-shaped gem with many triangular facets) in the physical worlds: (Ezekial Ch 1: 4-28 ; 2 Kings Ch 2:11). (The MerKaBah is explained in full detail in "The Keys of Enoch" by J.J. Hurtak). Literally MerKaBah means : "Mer" = Counter-Rotating Tetrahedronal Fields, "Ka" = Spirit & "Ba" = Body. The MerKaBah is thus a Space Time Vehicle that can take you inter-dimensionally. Your Aura is an intelligent field. With an understanding of Sacred Geometry, coupled with 18 Special Breaths, you can transcend Earthly Consciousness. This technique was taught by Drunvalo Melchizedek in "Flower of Life" workshops. The following page explains more of these secrets.

FLOWER OF LIFE WORKSHOPS,

originally taught by artist & teacher Drunvalo Melchizedek, employs Egyptian Mystery School knowledge & Earth's history (Lemuria etc) & teaches the ancient science of the MerKaBa (our Vehicle of Ascension) which has a scientifically measureable field of light extending for 55 feet around the body. Through an 18 breath meditation that takes only 4 minutes, this MerKaBah is REMEMBERED & re-activated, ending the deep, shared sense of spiritual separation. The symbols of the spiritual geometries, as shown in Figs 133b & 133c are essential triggers to assist in the regaining of our direct connection to Higher Self & ultimately to God. All other star—sun—planetary beings in this & other universes know about this MerKaBah space—time Vehicle, we Earthlings are the only ones who have forgotton it. We also have forgotton to Breathe. This Flower of Life workshop (write to me for more info) allows an ancient way of breathing to be re-instated, at the same time re-training the Pineal Gland (our own bio-inner Crystal) for its higher purpose. The Star Tetrahedronal fields (fig 133c) spinning within your Pranic fields (circle = sphere), helps to heal & restore lost memory.

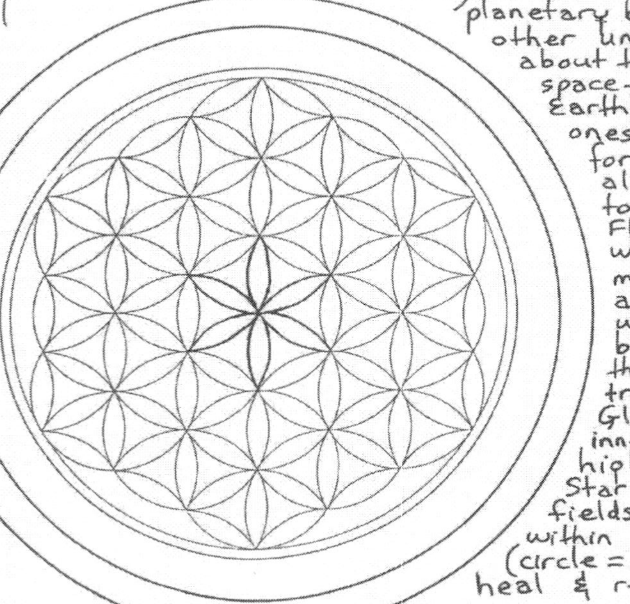

FIG 133b

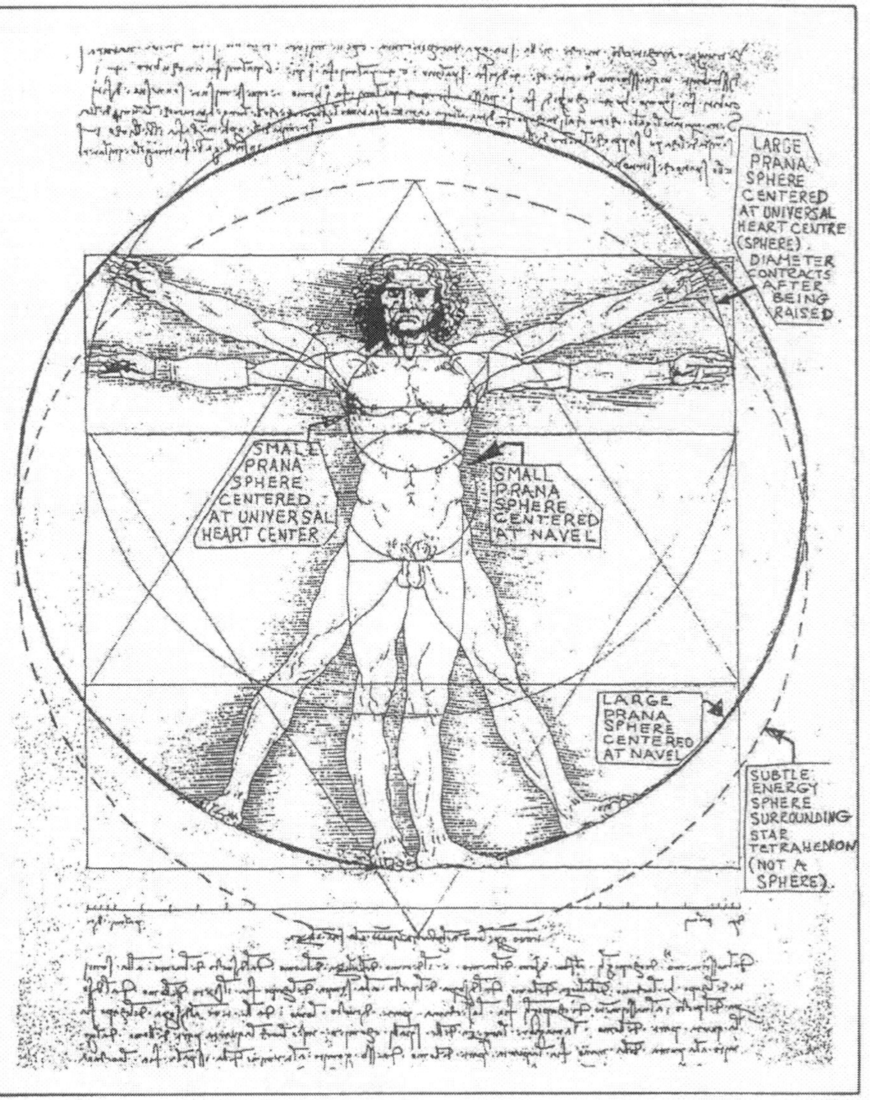

FIG 133c

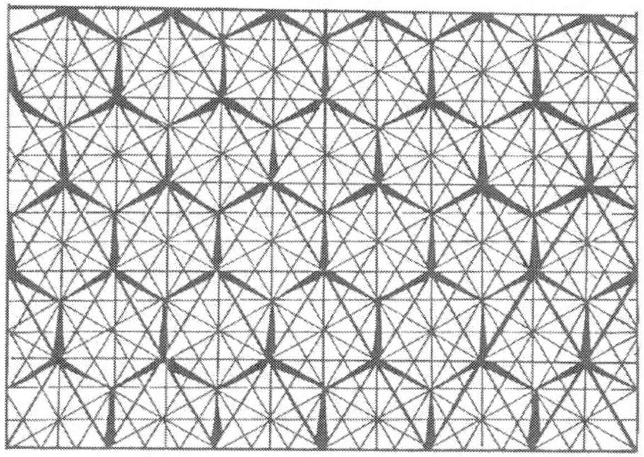

Fig 135

(Prayer of **TRIANGLES** an organization to link the world thru Meditation).

" In the centre of the Will of God I stand. Naught shall deflect my will from HimHer. I implement that will by love. I turn towards the field of service. I, the Triangle Divine, work out that will within the square & serve my fellowmen fellowomen".

Many people in a meditative or drug-induced state claim not only to see aurae of people but all living things & objects. Even further, they describe Space or our Atmosphere as an intricate rainbowic geometric pulsating WEB that can hold the Vibrations or the Thoughts of all Time of all the Ages past present future. To represent this, I offer Fig 135 above. It shows the Interference Pattern from multiple spherical sources filling space with an omni-triangular pattern of stable standing wave patterns. The two diagrams on the left show that every nodal point is a spherical source of 12 radiating vectors. These are equiangular & are equal in length.

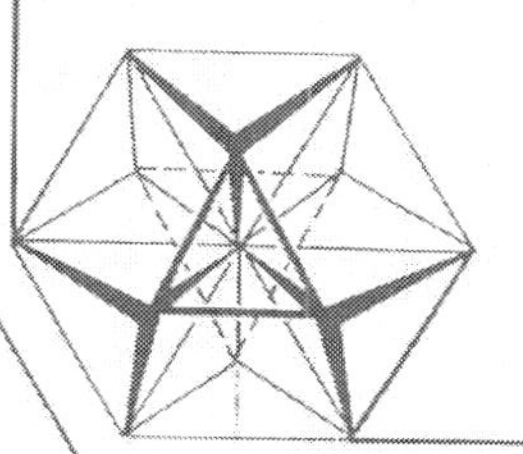

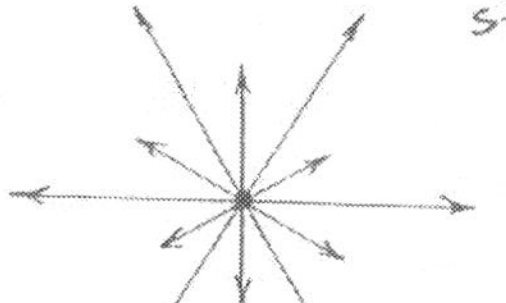

The Mystik Rose

Name of the Rose

Fig 136

BOOK BLESSING

By the One Power over all
By Diana, Hecate & Pan
Babaji, Iesus, Buddha & the Glory of You,
The Goddessence of You
Working for & thru this Majik Square
Paradigm Book
This book only reaches those for whom
it is intended;
This work cannot be mis-used;
And for those who use it rightly
For the good of all,
And according to free will
May all the Positive Majic of the Universe
Bring Joy & Power for good
That cannot be measured
And so mote it be.

You are NOW
Crossing over into Fourth Dimension
Of Health, Wealth, Love
& Perfect Self-Expression

Towards Majik Square Mandaloid Consciousness
Towards The RECRUDESCENCE of Majik Square Art.

TO CONCLUDE CHAPTER 12

PLEIADIAN TWELVE CHAKRA MEDITATION

(Our Solar System rotates about another greater Sun in the Pleiades star cluster. A Collective Consciousness from there channels thru a woman "Barbara Marciniak". We adults 'Children of the Light' have 2 helix in our DNA. Light carries information, data, & our children borned now "The Children of Love (Creation) have 12 helical strands of DNA evolving. Close your eyes & meditate. Have a bell rung to Awaken each chakra. The 12 DNA helixes relate to the 12 chakras :

CHAKRA No. **1** : Base, Survival. Awaken Kundalini serpent energy.
CHAKRA No. **2** : Sexual. Coming into full body love. Tantric sunrise, transmuting ejaculations up spine to nourish brain etc.
CHAKRA No. **3** : Solar Plexus. Getting in touch with your Emotions. 'Manipura' means 'Jewel filled'.
No. **4** : "Anahata Guru" My HEART is my Teacher.
No. **5** : Throat. Golden Mouth. Truth. Creativity.
No. **6** : 3rd Eye. Clear Vision. Clear Purpose.
No. **7** : 1000 petalled lotus. Fountainhead of Youth.
No. **8** : Just above the head. Say, "I AM MORE THAN THIS BODY". O Dreamer, Awaken !".
No. **9** : Beyond Earth's atmosphere i.e Moon region.
No. **10** : Edge of Solar System, extending further.
No. **11** : Edge of Galaxy, extending further, connecting.
No. **12** : Unifing with all Galaxies, with The ALL THAT IS. Linking with All Loving Life Forms. I Am God. God I Am, Awaken.

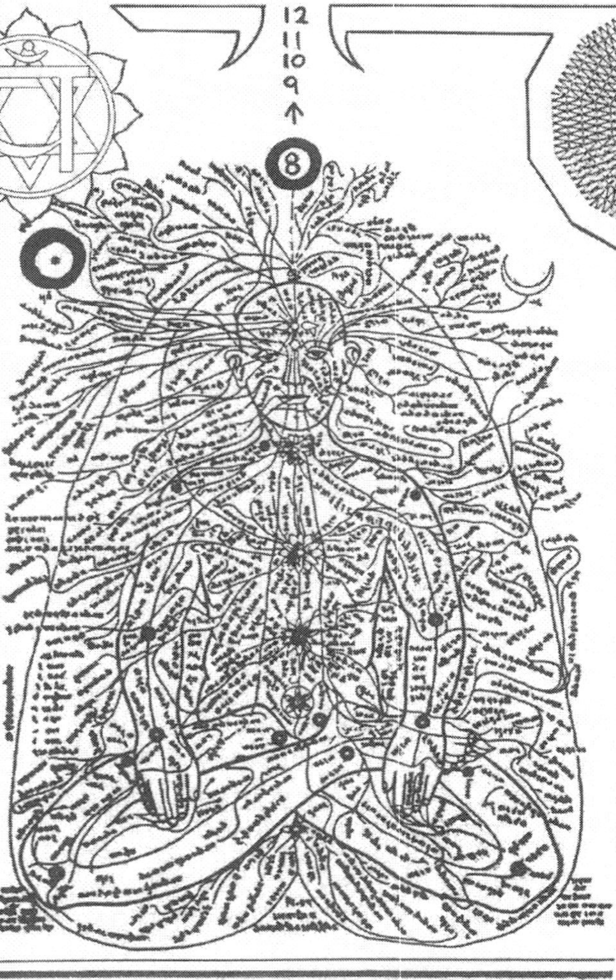

WE LIVE ON SEVERAL PLANES AT THE SAME TIME

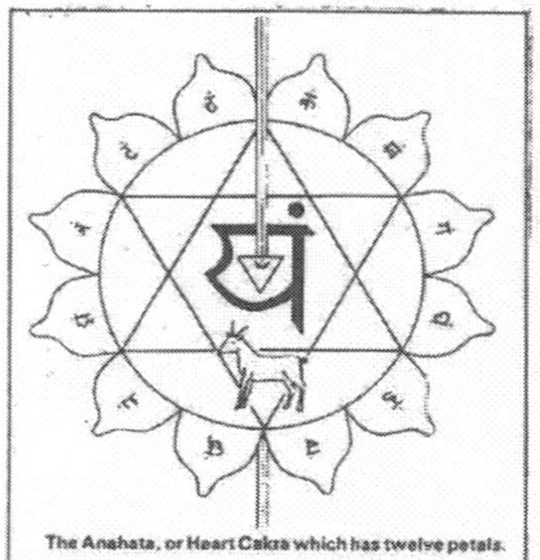
The Anahata, or Heart Cakra which has twelve petals.

introducing "the Benjamin Franklin Majik Square of 6teen (16)".

ROPERTIES of the 16 × 16 FRANKLIN SQUARE

PURITY
PEACE • PROSPERITY

RENEW YOURSELF

Body Soul Spirit

200	217	232	249	8	25	40	57	72	89	104	121	136	153	168	185
58	39	26	7	250	231	218	199	186	167	154	135	122	103	90	71
198	219	230	251	6	27	38	59	70	91	102	123	134	155	166	187
60	37	28	5	252	229	220	197	188	165	156	133	124	101	92	69
201	216	233	248	9	24	41	56	73	88	105	120	137	152	169	184
55	42	23	10	247	234	215	202	183	170	151	138	119	106	87	74
203	214	235	246	11	22	43	54	75	86	107	118	139	150	171	182
53	44	21	12	245	236	213	204	181	172	149	140	117	108	85	76
205	212	237	244	13	20	45	52	77	84	109	116	141	148	173	180
51	46	19	14	243	238	211	206	179	174	147	142	115	110	83	78
207	210	239	242	15	18	47	50	79	82	111	114	143	146	175	178
49	48	17	16	241	240	209	208	177	176	145	144	113	112	81	80
196	221	228	253	4	29	36	61	68	93	100	125	132	157	164	189
62	35	30	3	254	227	222	195	190	163	158	131	126	99	94	67
194	223	226	255	2	31	34	63	66	95	98	127	130	159	162	191
64	33	32	1	256	225	224	193	192	161	160	129	128	97	96	65

FRANKLIN 16×16 SQUARE.

fig 137

THE FRANKLIN SQUARE

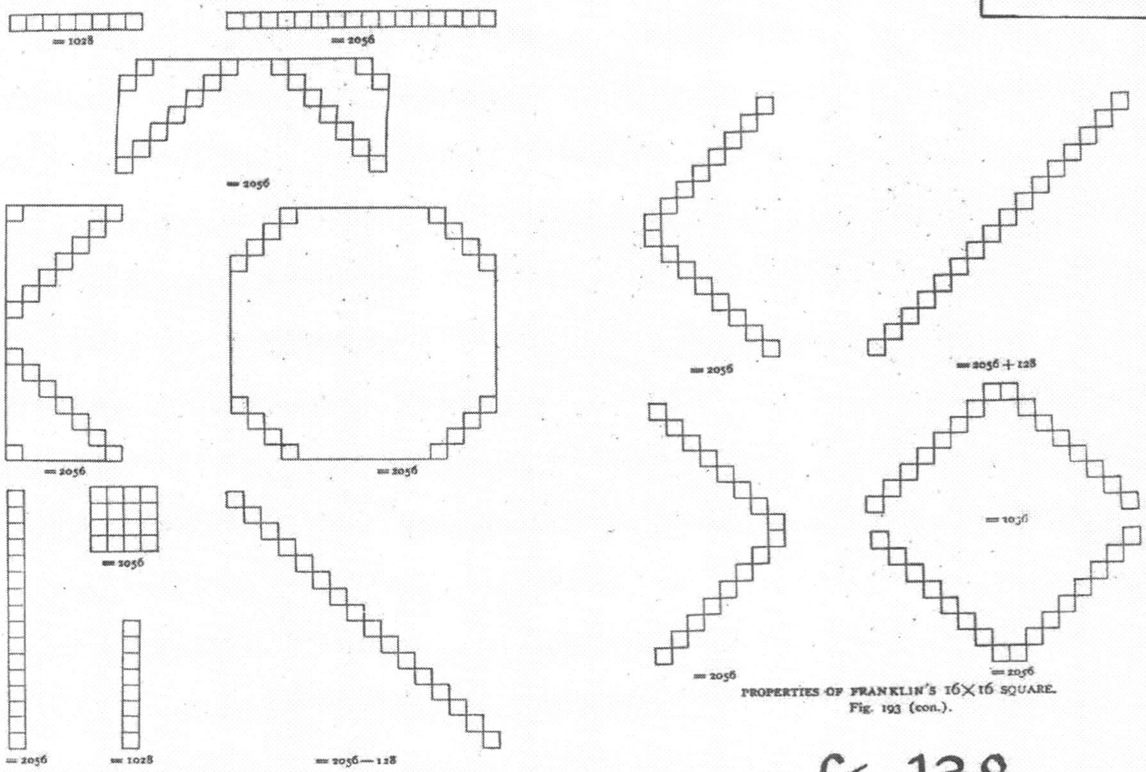

PROPERTIES OF FRANKLIN'S 16×16 SQUARE.
Fig. 193 (con.).

fig 138

Fig 139 →

Of the 256 Nos. of the Franklin 16 × 16 Square, an interesting pattern emerges with the first quarter or the Nos. from 1 → 64 as shown in Fig 139. Draw this pattern (that repeats) in the space of Fig 140.

				8	25	40	57								
58	39	26	7												
				6	27	38	59								
60	37	28	5												
				9	24	41	56								
55	42	23	10												
				11	22	43	54								
53	44	21	12												
				13	20	45	52								
51	46	19	14												
				15	18	47	50								
49	48	17	16												
				4	29	36	61								
62	35	30	3												
				2	31	34	63								
64	33	32	1												

FIG 140

Fig 137 →

All of the 16 columns, 16 rows & bent diagonals have a sum of 2056.
All 4x4 sub-cells = 2056.
All 2x4 sub-cells → have a sum of $\frac{1}{2}$(2056) = 1028.

200	217	232	249	8	25	40	57	72	89	104	121	136	153	168	185
58	39	26	7	250	231	218	199	186	167	154	135	122	103	90	71
198	219	230	251	6	27	38	59	70	91	102	123	134	155	166	187
60	37	28	5	252	229	220	197	188	165	156	133	124	101	92	69
201	216	233	248	9	24	41	56	73	88	105	120	137	152	169	184
55	42	23	10	247	234	215	202	183	170	151	138	119	106	87	74
203	214	235	246	11	22	43	54	75	86	107	118	139	150	171	182
53	44	21	12	245	236	213	204	181	172	149	140	117	108	85	76
205	212	237	244	13	20	45	52	77	84	109	116	141	148	173	180
51	46	19	14	243	238	211	206	179	174	147	142	115	110	83	78
207	210	239	242	15	18	47	50	79	82	111	114	143	146	175	178
49	48	17	16	241	240	209	208	177	176	145	144	113	112	81	80
196	221	228	253	4	29	36	61	68	93	100	125	132	157	164	189
62	35	30	3	254	227	222	195	190	163	158	131	126	99	94	67
194	223	226	255	2	31	34	63	66	95	98	127	130	159	162	191
64	33	32	1	256	225	224	193	192	161	160	129	128	97	96	65

In the sacred Birthing of New Consciousness Space in Fig 141 draw the B. Franklin 16 x 16 Yantra at 0° (of Fig 137). Observe the pattern at a $\frac{1}{4}$, $\frac{1}{2}$, $\frac{3}{4}$ progressions.

FIG 141

PURSUE YOUR PASSION

The message of the book is not to get into Majik Square Art but to get into your Art or Musik, to pursue your Passion, your Love.

Shown below is Fig 141 cut in half & joined end to end. Observe how this yantram looks like electrical oscillations or musical waves. (The Constant of the 16x16 Square is 2056. The intervals of the majik square Constants relates directly to the Intervals of Music Scales).

"Come, come whoever you are, Wanderer Worshipper, lover of leaving".

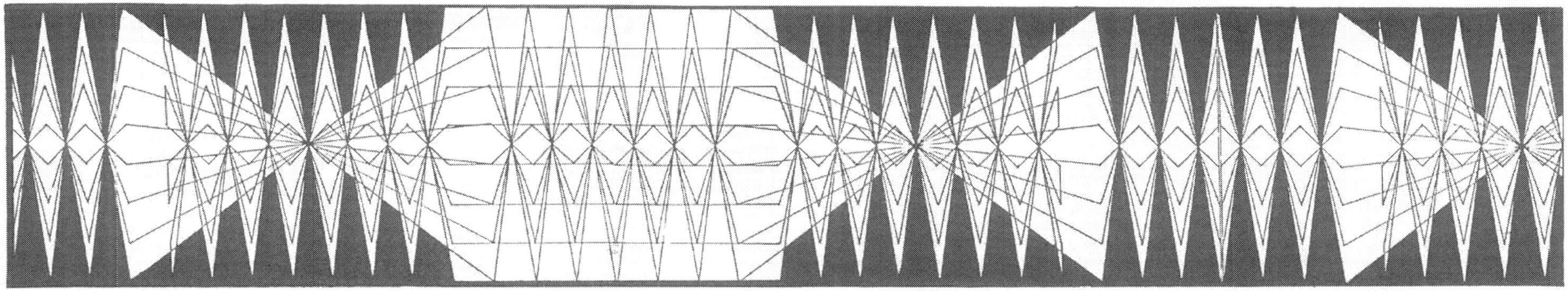

I have in the last 11 years painted 18 M.Sq. Banners in fabric paint upon cotton bedsheets 2m x 3m with edges braided & corners strung so that they can be exhibited in forests at festivals. Mainly, it is a Portable Art Form, able to take my Life's Work Overseas. This page shows a fotographic Progression from beginning to end. Top left: shows the Franklin 16 x 16 Square Yantra at ¾ Progression. The Rainbow petal's paints are mixed with flower essences & gem elixirs which I make myself.

I now manifest in my Life, the Funds necessary to manifest a full-color Production of all my artistic works. May this inspire you to pursue your Goals.

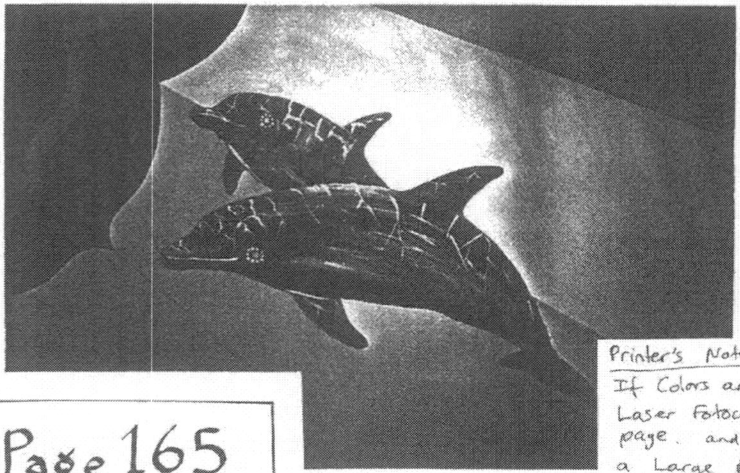

Printer's Note: If Colors are lost, Laser Fotocopy this page. and use a Large Dot Screen for better fotocopying.

Announcing a new and unique

MAGIC SQUARE OF 16 x 16

"BI - FOCCI"

COLUMN, ROW & DIAGONALS CONSTANT = 2056

1	255	243	13	5	251	247	**9**	**8**	250	246	12	4	254	242	**16**
240	**18**	30	228	236	22	26	**232**	**233**	23	27	229	237	19	**31**	225
33	223	**211**	45	37	219	215	**41**	**40**	218	214	44	36	**222**	210	48
208	50	62	**196**	204	54	58	**200**	**201**	55	59	197	**205**	51	63	193
65	191	179	77	**69**	187	183	**73**	**72**	186	182	**76**	68	190	178	80
176	82	94	164	172	**86**	90	**168**	**169**	87	**91**	165	173	83	95	161
97	159	147	109	101	155	**151**	**105**	**104**	**154**	150	108	100	158	146	112
144	114	126	132	140	118	122	**136**	**137**	119	123	133	141	115	127	**129**
128	130	142	116	124	134	138	**120**	**121**	135	139	117	125	131	143	113
145	111	99	157	149	107	**103**	**153**	**152**	**106**	102	156	148	110	98	160
96	162	174	84	92	**166**	170	**88**	**89**	167	**171**	85	93	163	175	81
177	79	67	189	**181**	75	71	**185**	**184**	74	70	**188**	180	78	66	192
64	194	206	**52**	60	198	202	**56**	**57**	199	203	53	**61**	195	207	49
209	47	**35**	221	213	43	39	**217**	**216**	42	38	220	212	**46**	34	224
32	**226**	238	20	28	230	234	**24**	**25**	231	235	21	29	227	**239**	17
241	15	3	253	245	11	7	**249**	**248**	10	6	252	244	14	2	**256**

Callum Coats, April 1995

Fig 142

"Imagination is more important than knowledge" ... Albert Einstein

The M.Sq. of Fig 142 has its 0° Yantram (Fig 143) already constructed for you. The reason why you need not draw it is that it is very similar to the one you just drew in Chap 12 known as Fig 123. (Refer to your "Solutions" to compare figs 124 & 143 (i.e. 12×12 & 16×16).

2 The Golden Flower represented as the most splendid of all flowers.

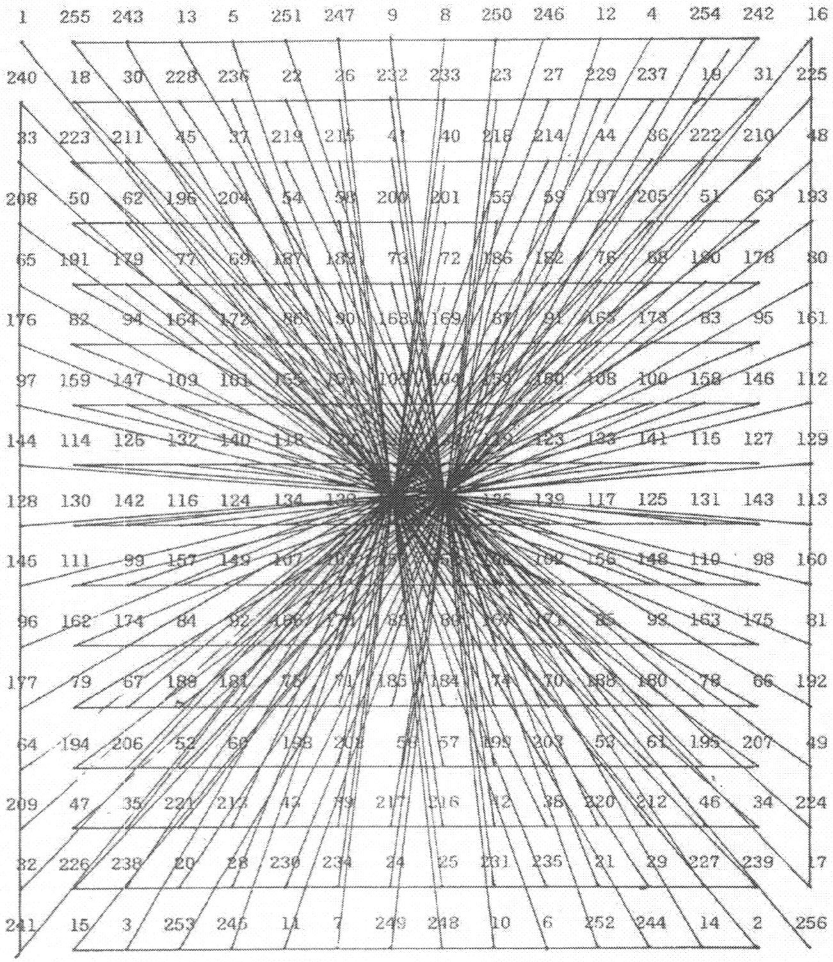

Fig 143

SEQUENTIAL CONNECTIVE LINE DIAGRAM
FOR 16 x 16 MAGIC SQUARE

Page 168

fig 144

TRATAKA

Have the yantram of Fig 144 (a 16-pointed star derived from the Majik Square of 3 at 8 × 22·5°) approximately 1 metre away from your eyes & stare at it whilst in a meditational posture. Mindlessly gaze at its centre for 3 minutes then close your eyes. Observe the AFTER-IMAGE. Then repeat the process, stare at the yantram again, without blinking for 3 minutes & close your eyes again, focussing your consciousness into your inner Eye, into the after-image, to the extent that awareness of the rest of the body is lost. What color does the Red background change to ? Is it the Complementary Opposite Color as known in the science of the Color Wheel ?

"TRATAKA" literally means "a concentrated form of Gazing. It is an ancient Indian Hatha Yoga technique for controlling a temptestuous mind & its thought waves. Regular practice of Trataka helps to open the door of your Inner or Third Eye. Trataka can be practised on The Rising Sun or Full Moon, on one's own image in a mirror, in the eyes of your lover, photographs of your spiritual masters/mistresses etc. MySelf, I find much peace & insight when I focus my gaze on a certain majik Square Yantram's centrepoint or "Bindu". The after-images are also very pacific. At forest exhibitions, it is quite common to see people staring at these majik square banners (2m × 3m), overwhelmed by their size, color & peacefulness.

The diagrams below, Fig 145 show the EVOLUTION of the Lo-Shu that forms Fig 144 :

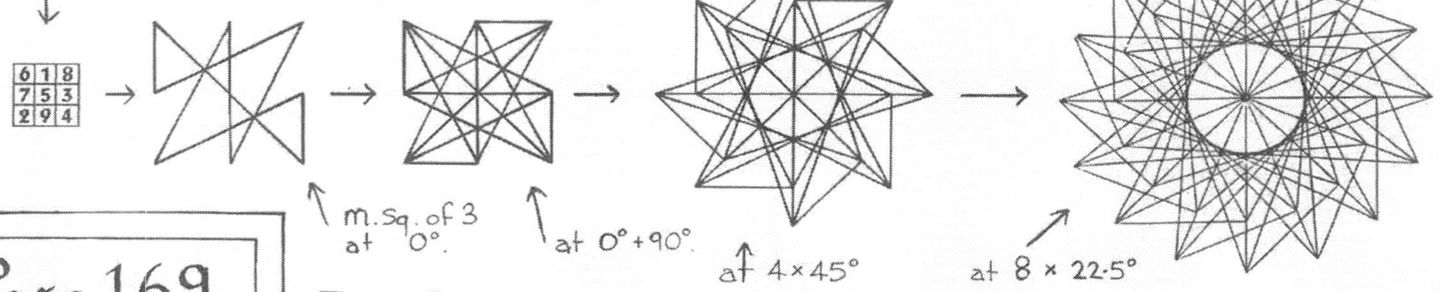

FIG 145

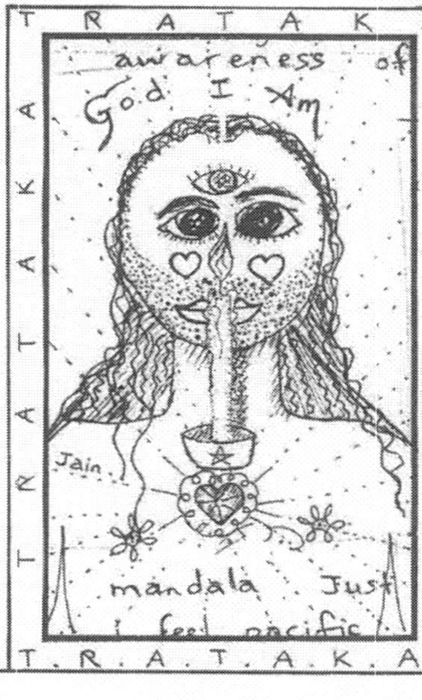

Page 171

A Request

I feel tempted to meet You, You who are presently occupied by reading this book. To dissolve this distance & mystery between author & reader, can we meet thru the interface of a certain Majik Square? that You will draw (or manifest on your computer) & one that I have never seen & will be charmed by. On this page are 16 x 16 Grid of dots. I would like to see what the B. Franklin M.Sq. of 16 x 16 looks like at 0° + 90° + 180° + 270° ie. at 4 x 90°. (it requires over a 1000 lines to be drawn, but probably 500 repeat themselves & won't have to be drawn). The result of such an effort is always a Surprise. I could do it, or on a friend's computer, but I never have yet, guess I'm too busy with manuscripts, murals, children, gardens, relationships, dance performances etc & other necessary illusions.

So I invite you to draw this Never-Seen-Before Crystal Star, color it in with jewel-filled affirmations & write to Box 396 c/o P.O Byron Bay 2481, N.S.W, so we can meet. I don't know why I am writing this, but it will become apparent when we connect.

Your design will become a large circular Decal.

"To wake at dawn with a winged Heart
give thanks for another day
of Loving.
To rest at the noon hour
meditate love's ecstasy."
(Kahlil Gibran).
Jain.

Page 171 Ā

Shown on this page is The Benjamin Franklin Magic Square of 16 × 16 Yantra (i.e Pattern Fig 141 drawn on p 163) superimposed upon itself at 4 × 90°.
This diagram is known as Fig 141 Ā.

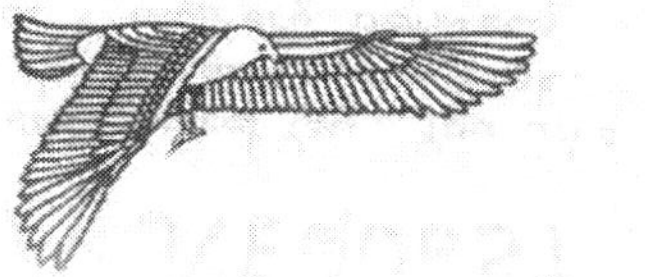

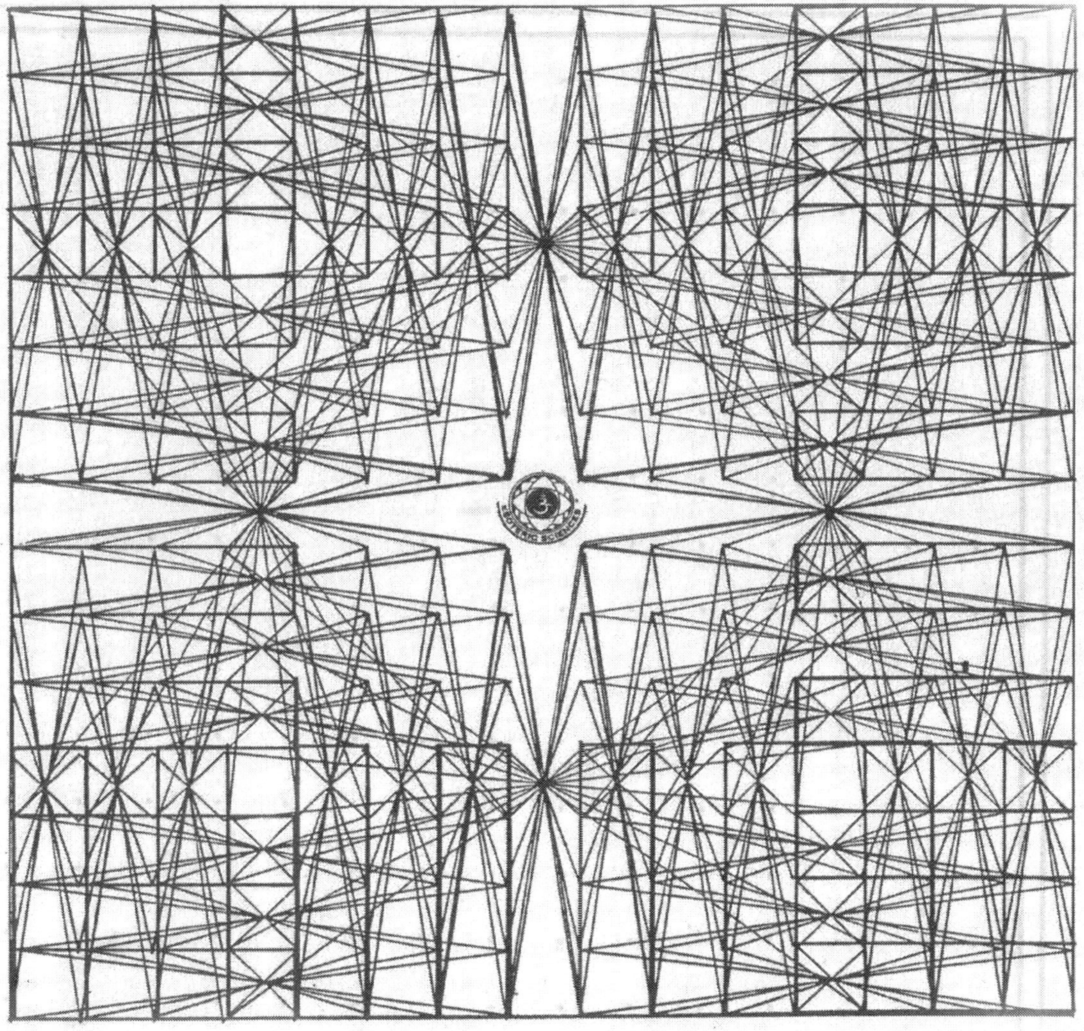

Fig 141 Ā

The Benjamin Franklin Magic Square of 16 × 16 Yantra at 4 × 90° (Fig 141A) tiled or tessellated 4 times to form an intricate pattern known here as Fig 141 B.

Page 171 B

Fig 141 B

Page 171C

The Benjamin Franklin Magic Square of 16 × 16 at 4 × 90° (Fig 141 A) tiled or tessellated 16 times. (This may be written in shorthand as "The B. Franklin M. Sq. 16 × 16 at 4 × 90° T × 16").

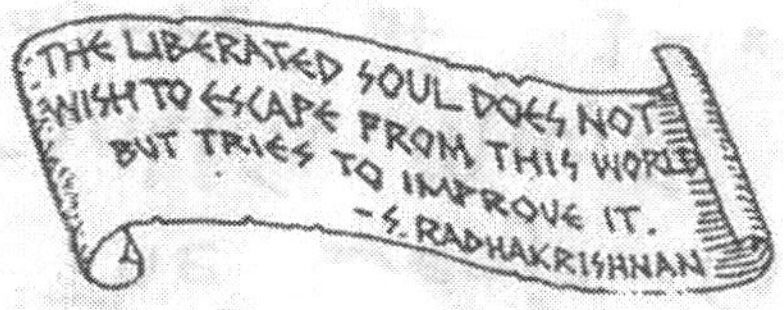

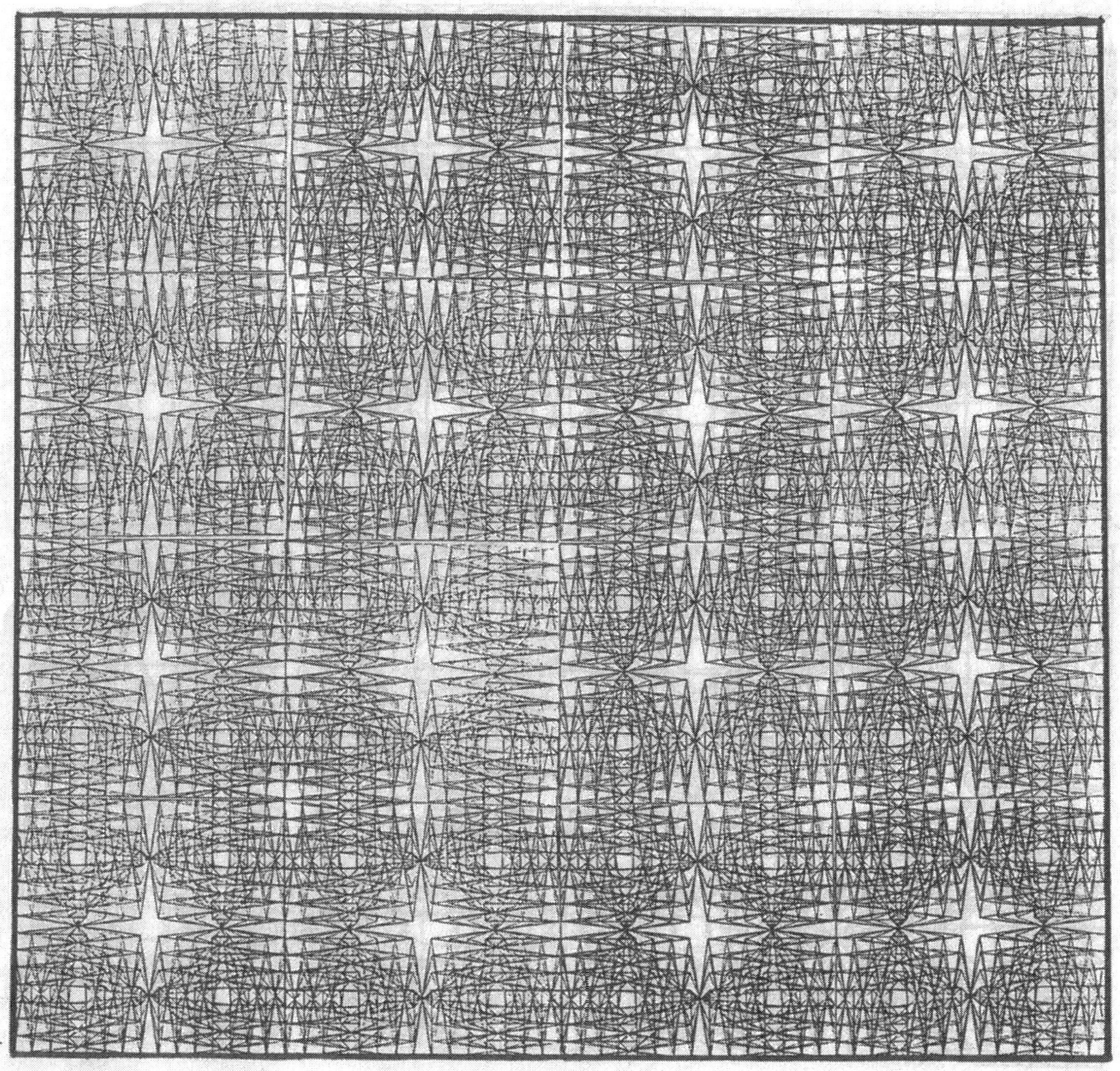

Fig 171C

Border and row ornaments, Peter Behrens, 1908

And so the time comes when all the people of the earth
can bring their gifts to the fire
and look into each other's faces
unafraid

Breathe deep
Feel the sacred well that is your own breath, and look
look at that circle
See us come from every direction
from the four quarters of the earth
See the lines that stretch to the horizon
the procession, the gifts borne
see us feed the fire
Feel the earth's life renewed
And the circle is complete again
and the medicine wheel is formed anew
and the knowledge within each one of us
made whole

Look around the circle, look at our faces
Each one different, each special

We struggled
We held out our hands and touched each other
We remembered to laugh
We went to endless meetings
We said no
We put our bodies on the line
We said yes
We invented, we created
We walked straight through our fears
We formed the circle
We danced

We spoke the truth
We dared to live it

From: Truth or Dare - Starhawk (c)

Page 171 D

CHAPTER: X

INTRODUCING THE VISUAL MULTIPLICATION-TABLE FROM IT'S REDUCTION GRID.

INTRODUCING the VISUAL MULTIPLICATION-TABLE and its REDUCTION-GRID of ONE & 8 (of ONENESS).

×	1	2	3	4	5	6	7	8	9
1	1	2	3	4	5	6	7	8	9
2	2	4	6	8	10	12	14	16	18
3	3	6	9	12	15	18	21	24	27
4	4	8	12	16	20	24	28	32	36
5	5	10	15	20	25	30	35	40	45
6	6	12	18	24	30	36	42	48	54
7	7	14	21	28	35	42	49	56	63
8	8	16	24	32	40	48	56	64	72
9	9	18	27	36	45	54	63	72	81

FIG 146

1	2	3	4	5	6	7	8	9
2	4	6	8	1	3	5	7	9
3	6	9	3	6	9	3	6	9
4	8	3	7	2	6	1	5	9
5	1	6	2	7	3	8	4	9
6	3	9	6	3	9	6	3	9
7	5	3	1	8	6	4	2	9
8	7	6	5	4	3	2	1	9
9	9	9	9	9	9	9	9	9

FIG 147

NB: Please Read P102, Now.

Fig 146 is the well-known Table of Multiplication. This can be easily transformed into the Reduction Grid of Fig 147. Fig 146 includes 2 digit numbers whereas Fig 147 has only single digits. A two-digited number can be reduced to a single digit by a process of addition e.g: 10 = 1 + 0 = 1 ; 81 = 8+1 = 9.

Fig 148 shows how every number of Fig 147 (The Reduction Grid) can be represented by a point or dot in the centre of each number's square. Take away the 9x9 frame & what is left is a matrix or lattice of dots or grid of 81 dots shown in Fig 148a.

Now we can begin to transform Number into Art. Firstly, we locate, in Fig 147 all the cells that have the number "1" written within them. Fig 149 shows all the cells of "One". Every cell is then connected to every other cell as shown in Fig 149a These connective lines of energy form a symbolic double-end or double-pointed healing pyramidal Crystal which has the Vibration of Oneness. In the space of Fig 151 you will repeat this diagram of Fig 149a. Fig 150 shows all the cells of "8" & Fig 150a shows its Connective Lines of Energy Yantram which interestingly is a 90° Phase Shift of Fig 149a. In the Space of Fig 152 you will repeat this diagram of Fig 150a. You may meditate upon the crystalline Yantram of Oneness as "The Higher & Lower Tetrahedrons of our Selves".

Then, in the space of Fig 153 you will combine the Yantram of Oneness with the Yantram of Eightness, i.e Figs 151 + 152 respectively. This is known as "The VISUAL MULTIPLICATION TABLE & its REDUCTION GRID of 1 (ONE) + 8 (EIGHT).

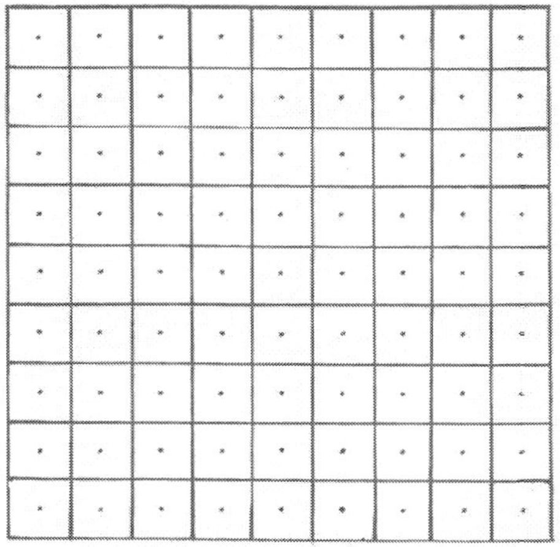

Fig 148

Fig 148a

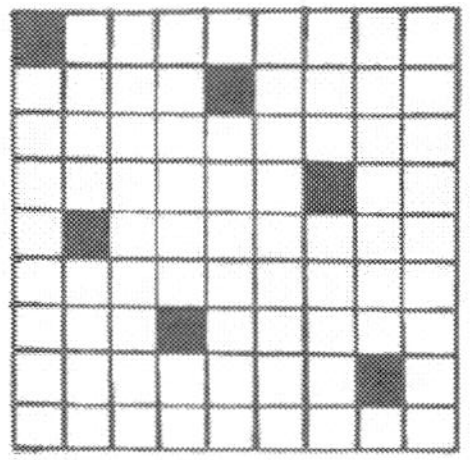

Fig 149

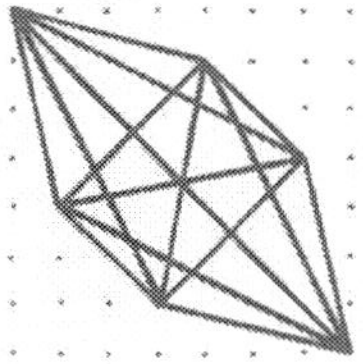

Fig 149a

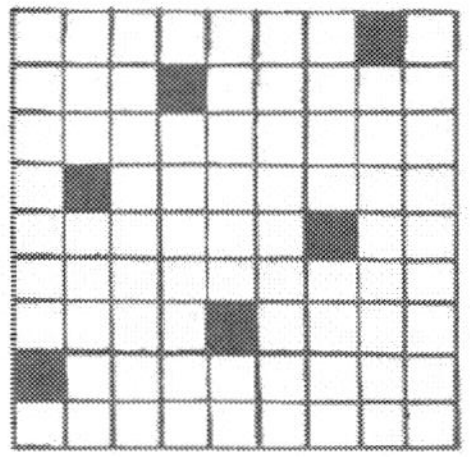

Fig 150

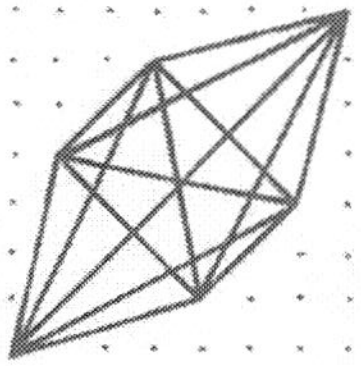

Fig 150a

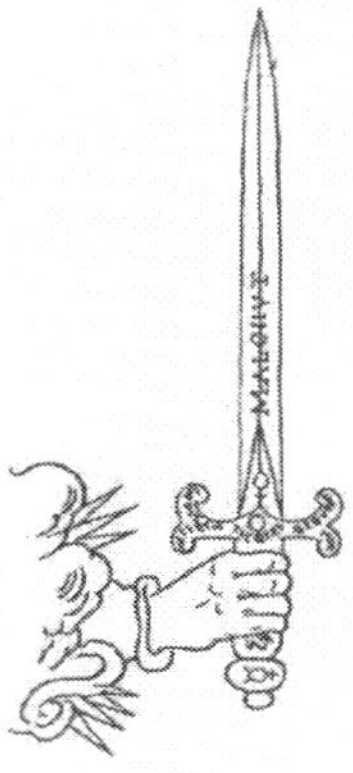

Fig 151

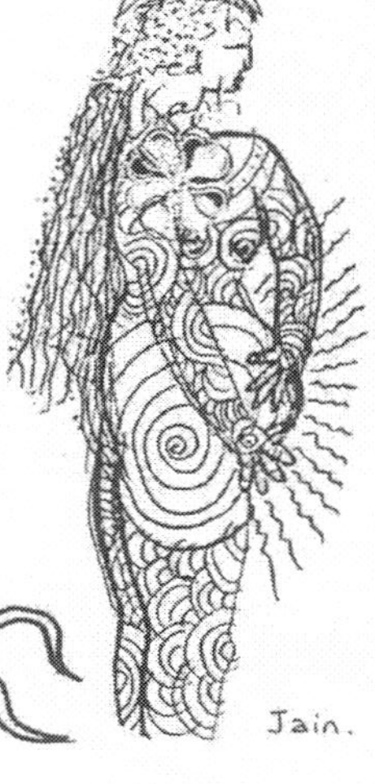

Fig 152

Fig 153

PLATINUM CRYSTAL

Fig 155

CREATES ATOMIC ART of Rutile + Platinum + BERYLLIUM

Isn't it interesting that what you just created in figs 151 + 152 is the Atomic Art Octahedral Shape of RUTILE (a sand that gets mined. It occurs on beaches as a black colouring). This is shown as Fig 154, a diagram that appears commonly in high school science textbooks.

Isn't it interesting that what you just created in fig 153 is identical to fig 155. Shown here on your left is a rare photo of PLATINUM CRYSTAL Hemisphere with a thousand crystal facets, many of them resolved into single atoms. These refraction photos are the closest visualization that science can give with respect to the nature of atomic substance, which appears to be patterns of geometrized light energy. Magnified 750,000 X by the late Erwin. H. Mueller, Penn State University. I have reprinted this atomic art from "THE KEYS OF ENOCH" by J.J. Hurtak. Quote: "If you look at platinum crystal under a field ion microscope, you see bubble formations which form pyramid shapes of light going from every stage of geometry all the way through the immediate field of the crystal. If you look at blood crystals under an electron microscope, you can find the shape of the pyramid field in the crystal forms of blood ... the Pyramid of Light is the central geometric form (in the macro & micro worlds)... A closer examination of the pyramidal units in Hydrogen atoms will further reveal the geometry of a STAR of SOLOMON as a life-giving form ..."

Continued

TITANIUM DIOXIDE = RUTILE

(Ionic Lattice Structure)

TiO_2 (rutile) structure.

Co-ordination 6:3.
The metal ion is surrounded by six non-metal ions at the corners of an octahedron; and each non-metal ion is surrounded by three metal ions at the corners of an equilateral triangle.
Examples:- TiO_2, MnO_2, MgF_2.

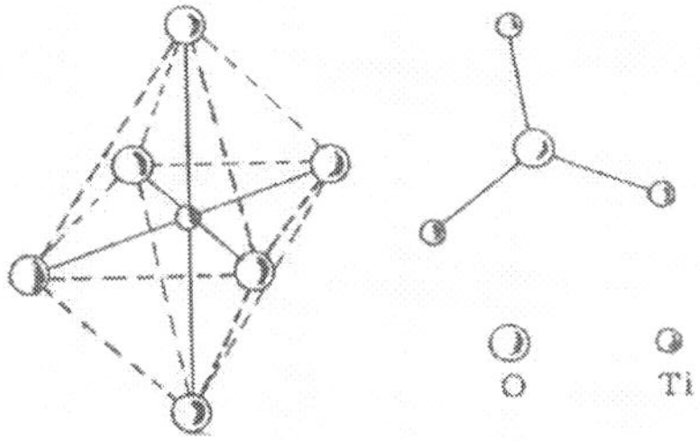

Fig 154

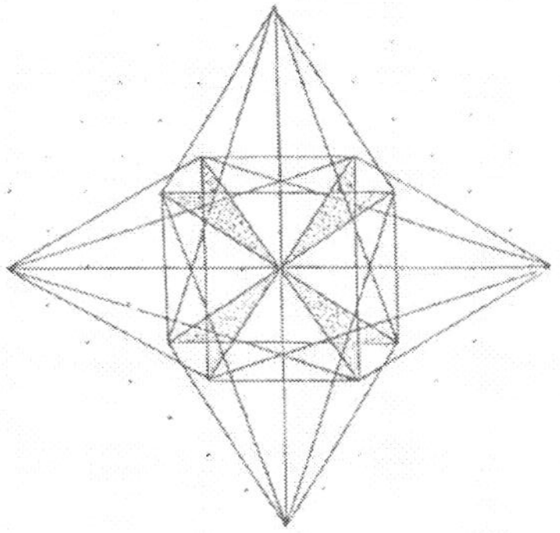

fig 153

BERYLLIUM

ATOMIC ART cont...

The invention of the Scanning Tunnelling Electron Microscope, in the 1980s, has equated new perspectives of ATOMSCAPES with sacred mathematics of the HUMAN BIO-PYRAMID. It has created the equations:

Fig 153 = Fig 155
Fig 153 = Fig 157 (Beryllium).

To understand this, it is important to visualize any 2-dimensional Yantram like Fig 153 as a 3-Dimensional Hologram (Fig 157).

Atomic Mandalaic Art is a new (ancient) art-form to help provide an enlightening change in the way we think, see & experience this world. Such designs are built-in secretly in the dome roof of the Alhambra Palace, Spain, plus in many tiled domes in Arabia.

The Octahedral form of Beryllium (Fig 157) comes from Chris ILLERT's book "Alchemy Today", which is a centennial commemoration of the two occult chemists Annie Besant + C.W. Leadbeater. Our modern nuclear geometries which are governed by the 5 Platonic Solids was first elucidated by Plato in his "TIMAEUS", then first drawn by the mentioned couple. One would enter the World of the Micro (siddha Yoga technique) & describe the clairvoyantly diagnosed geometries which allowed the other partner to draw them in detail. Buckminster Fuller, who developed the modern dome geometries adds further validity & verification to the work of Besant + Leadbeater. Basically, the tetrahedron is the basic building block of the Universe & such revelations has made obsolete the quantum physics of modern "particle physics": that they've got it all wrong!

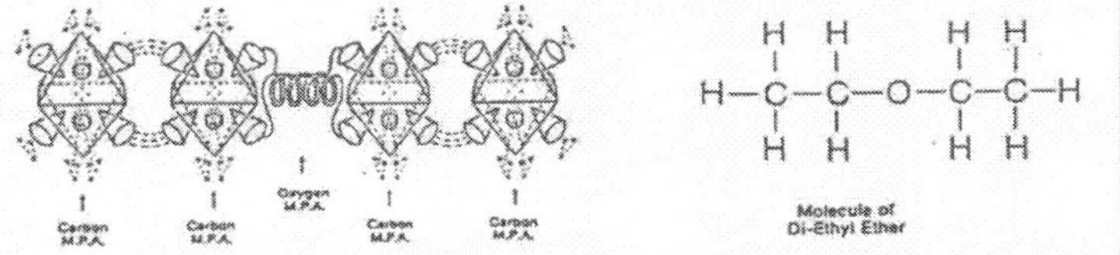

early "micro-psi" image — modern chemical formula

THE OCCULT CHEMISTS.

^{10}BERYLLIUM ground-state

$r = 2.46 \pm 0.05$ fm.

$\tau = 2.6 * 10^{6}$ years

mirror to ^{10}Carbon

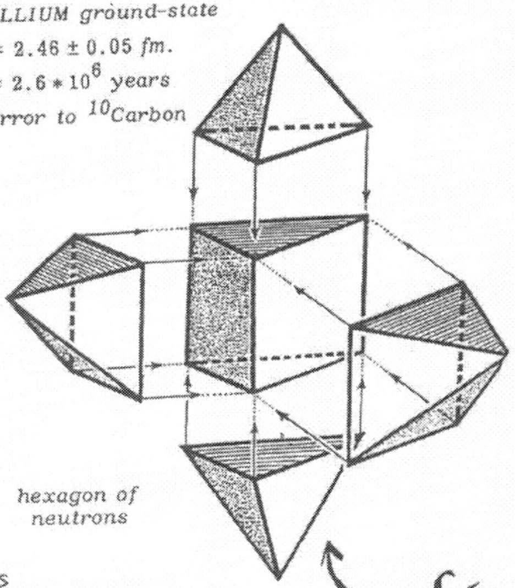

hexagon of neutrons

Fig 157

TITANIUM

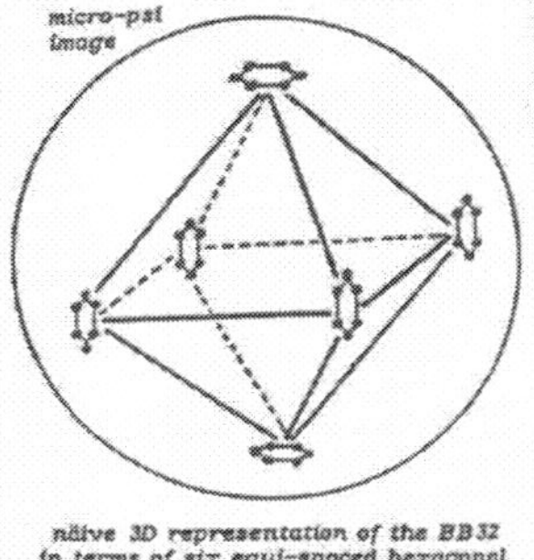

naive 3D representation of the BB32 in terms of six equi-spaced hexagonal facets, at the vertices of an octahedron, on the surface of a great sphere. Titanium was specifically placed in the "Octahedron group A" by early micro-psi observers: see "Occult Chemistry" (1952) page 223. This BB32 also occurs in the 5th main nuclear shell in elements such as ^{88}Strontium (82.5%).

Fig 156

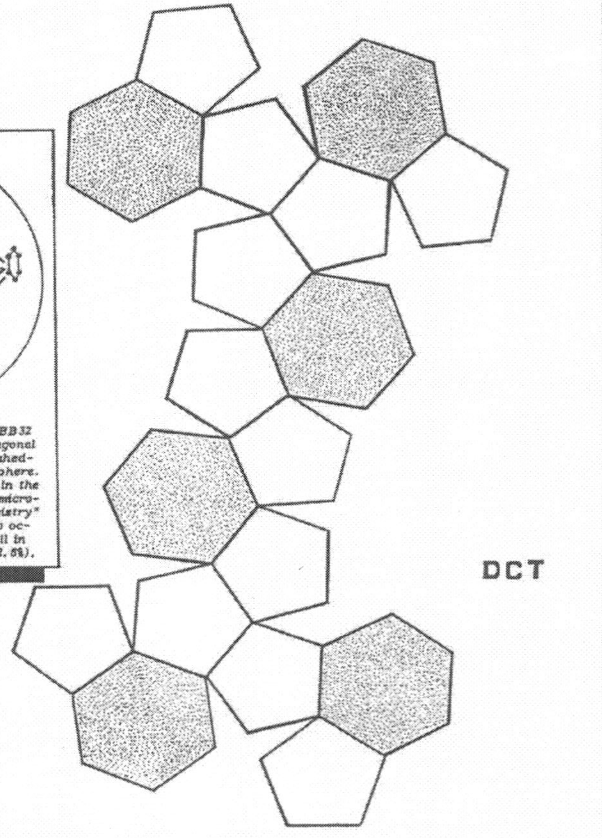

DCT

Pattern for making Buckey-Ball 32, as in the ^{48}Titanium outer shell. It has 32 vertices (nucleons), 12 pentagonal faces and 6 hexagonal faces, representing a union of 8 α-particles from the "1d", "2s" and inner "1f" subshells.

A COLLECTION OF OCTAHEDRAL SHAPES

16	2	3	13
5	11	10	8
9	7	6	12
4	14	15	1

This study is based on Fig 151 which you have just drawn. Fig 158 shows 9 m. Sqs. of 4 at 0°+90° tiled into a larger square. It shows the background (negative) space forming 12 octahedral shapes (all of which have been shaded-in.

FIG158

Two solids are considered to be duals if the number of vertices of one equals the number of faces of the other and they have an equal number of edges. The dual of a solid can be generated by inscribing a sphere so that it is tangent to the solid at the midpoints of its edges and then by drawing lines perpendicular to the edges and tangent to the sphere at those points.

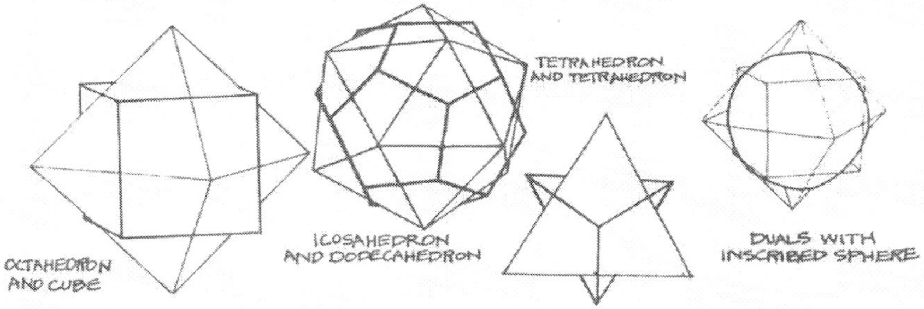

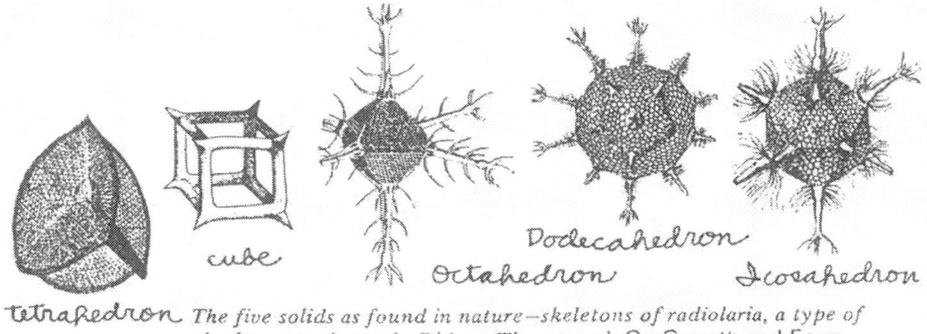

The five solids as found in nature—skeletons of radiolaria, a type of plankton, as shown in D'Arcy Thompson's On Growth and Form, *from left to right: Callimitra agnesae, Lithodubus geometricus, Circoporus octahedrus, Circorrhegma dodecahedra, Circognia icosahedra*

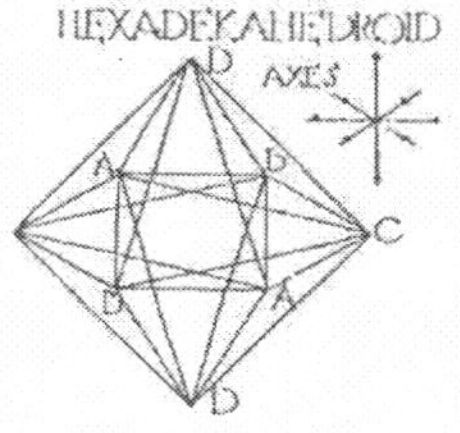

More Octahedral Shapes relating to the "X" of 1 double-ended Healing Crystal of Figs. 149a & 151.

The title "The Visual Multiplication Table from its Reduction Grid of One" can be abbreviated to '"X" of 1'. The appropriate symbol here is "X" which means "Multiplication" which means "Growth". Thus these Yantras of Chapter "X" represent <u>archetypes of Growth</u> & fertility.

This page & the previous one are presented merely to inspire you to view the Oneness of all Creation/the Octahedral Atomic Artforms.

This "Book of Magic Squares" is presented as a "THEOLOGUE" ('Theo' = God, 'Logue' = discussion') a talking with 'The God I Am, The God You Are.'

'Cos God is Life
I AM Life.
All mathematics
emanates from

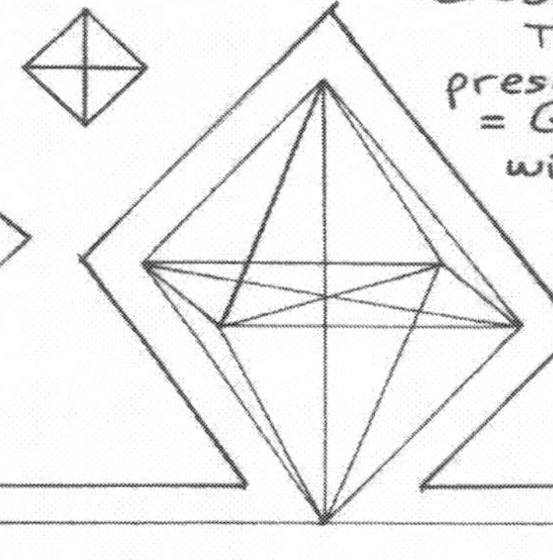

THE NUMBER ONE

The KNIGHTS TEMPLARS wore the Cross pointing to the four quarters in token of the Power by which a man may rise to loftier consciousness & which is found at the root of every form in the Universe.

Illustrated here is the Cross & the Cipher ⟶ ("Cipher" = a secret method of writing as by a specially formed code of symbols). The Cipher writing is reputedly derived from the component parts of the eight-pointed Cross. It is really a most ingenious device & yet simple, the key to which each Knight wore upon his breast. This is shown in Fig 159. The first four diagrams show how the Cipher was built up. The fifth "V" shows that "passing thru the Octagon" was "standing on the Cross".

The 8ight-pointed Cross is illustrated in this Chapter "X" 'cos this shape appears commonly: (Fig 153 at 0°+45° shown here on the top Right-hand side of this page = Fig 160). This 'Maltese Cross' also appears prevalently in other Majik Square Art centerpieces.

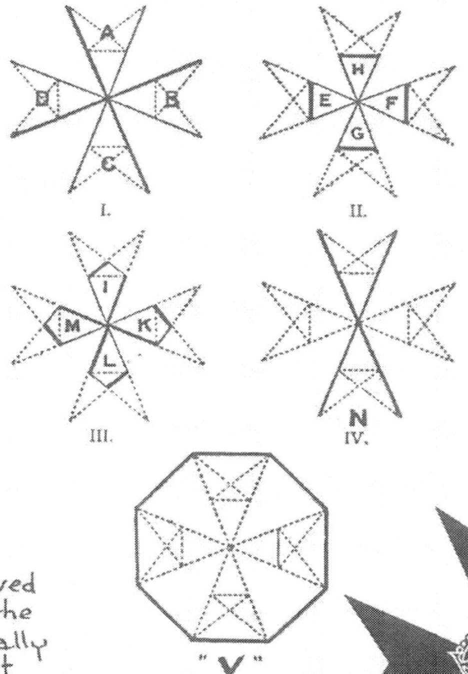

Fig 159

"THE BEST WAY TO LEARN ANYTHING IS TO DISCOVER IT BY YOURSELF."

Shown above is Fig 160 which you will draw next. Known also as The "X" of 1 (Fig 151) drawn 4 Times at 45° angles Yantram.

Around the banner I have painted the word "EARTH" in a necklace 14 times, so that now it reads "HEART HEART HEART" etc. It can also read : "HEAR THE ART" etc. This was originally painted on a banner 11 years ago. ©.

The Reduction Grid of Fig 147 has all its cells of "nines" in the top row & extreme right-hand-side-column. If we ignore this row & column of nines, we change the Reduction Grid from a 9x9 Lattice of Dots to an 8x8 (Fig 147a).

The challenge on this page is to draw the Yantram of Fig 153 at 0°+45°. To do this you will need to study Fig 160a: a square over-laid by its diamond (a 45° turn). Each square has 49 cells, i.e 7x7 which provide the grid of dots for the required 8x8.

Fig 160 is the space in which you will create the "X" of 1+8 at 0° + 45°. All the points of "1" & "8" have been given, all you have to do is locate which dots represent "1" or "8". Observe how all these points lie upon two circles! If you are not sure how to proceed compare Fig 160 to Fig 160a. In fine pencil, outline the square & its diamond in the space of Fig 160. Like usual, check your yantram that it is similar to the one shown in the back of this book (or on the previous page).

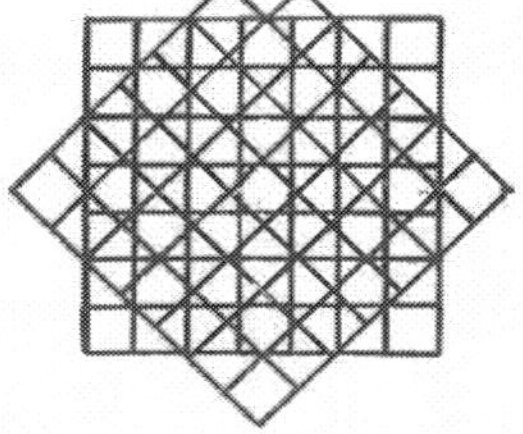

Fig 160a

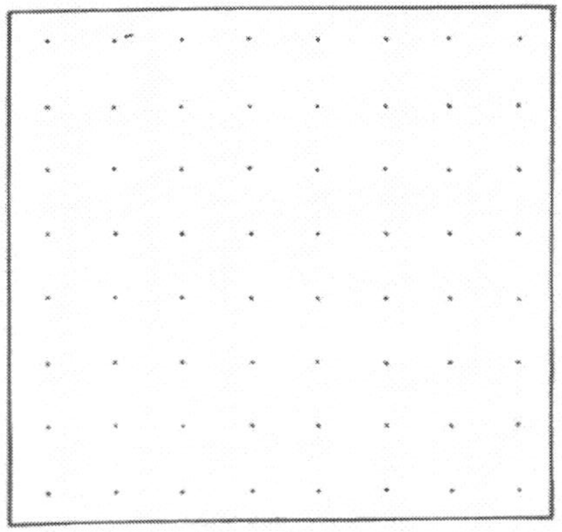

Fig 147a

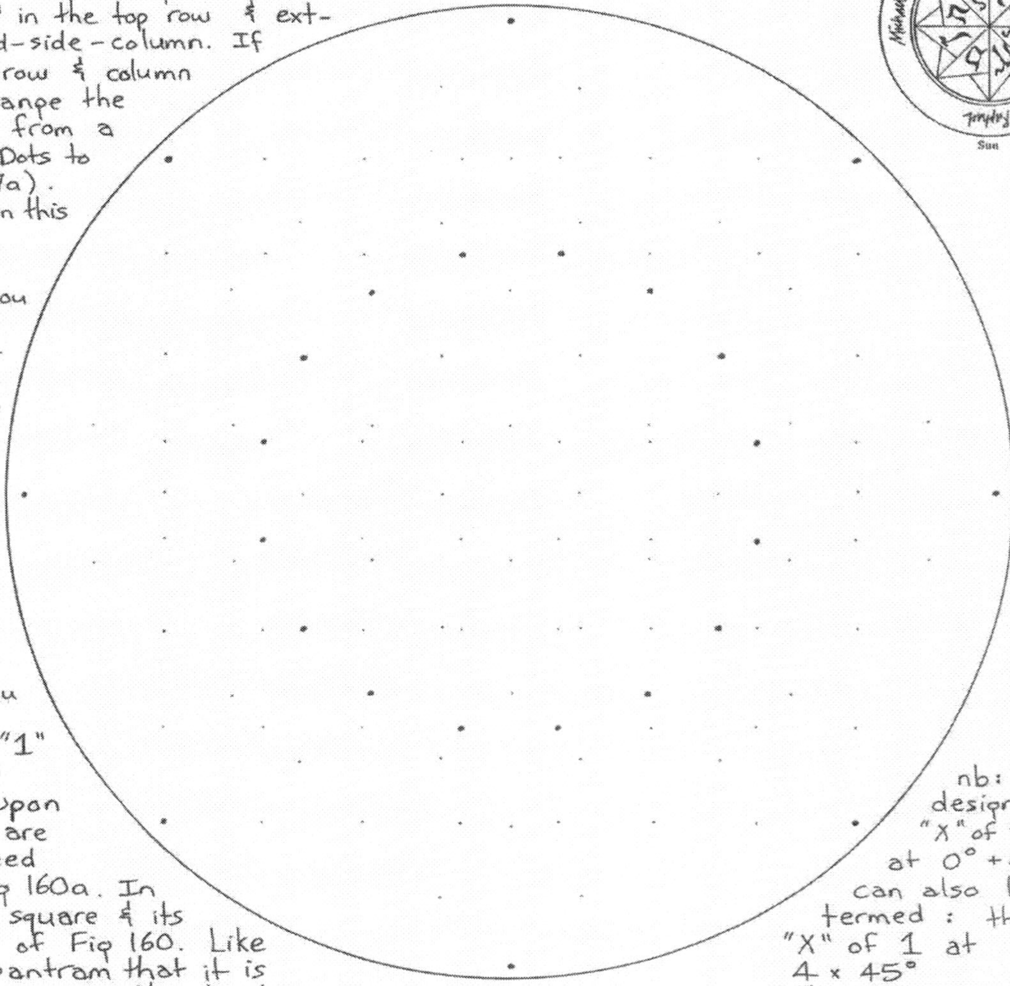

Fig 160

nb: This design "X" of 1+8 at 0° + 45° can also be termed: the "X" of 1 at 4 x 45° YANTRAM.

Jain.

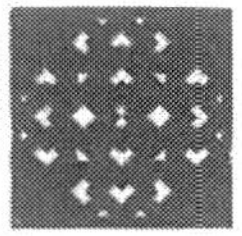

THE 'MAJIK SQUARE GALLERY

Fig 161 is the "X" of 1 (Figs 149a & 151) drawn 16 times at rotation angles of 11·25°
ie: Fig 161 = "X" of 1 at 16 x 11·25°.
It is also known as "X" of 1 + 8 (Fig 153) at 8 x 22·5°.

Fig 161

Exists for the dissemination of meaniful & sacred Geometry for our Newborns

Jain

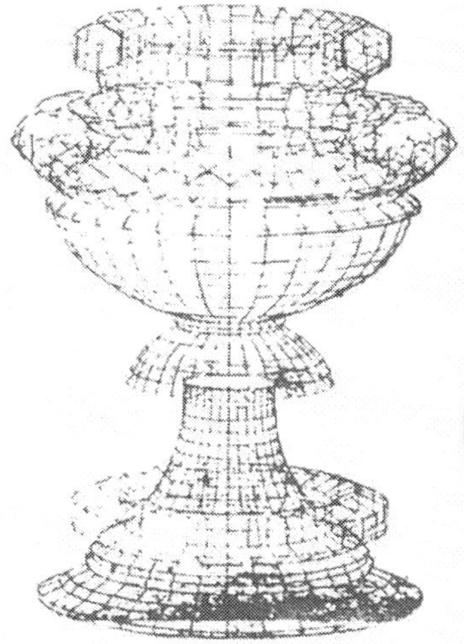

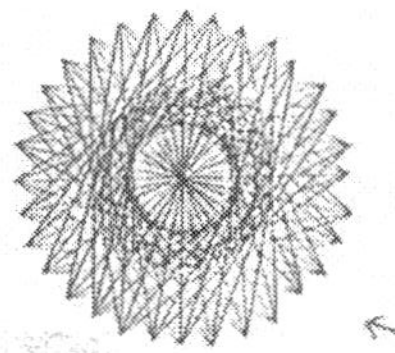

TRANSCENDENTAL

CONTEMPORARY

DESIGNERS

Fig 161.

Introducing the Multiplication Table and its Reduction Grid of Two (2) and Seven (7).

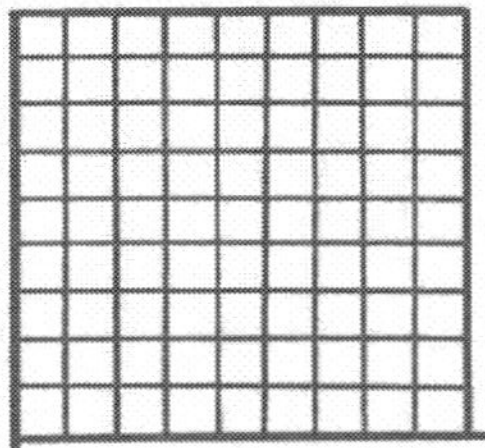

Fig 162

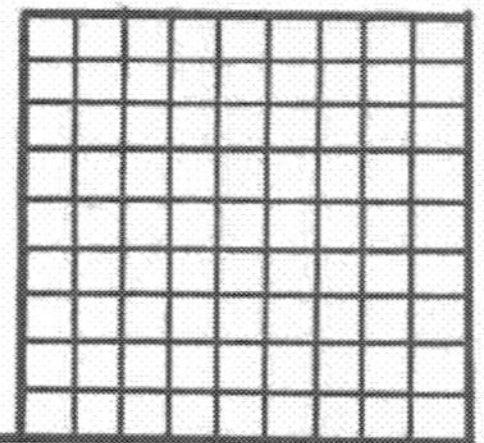

Fig 163

Fig 164

Fig 165

To complete Figs 162 & 163 refer back to the Reduction Grid of Fig 147. Then, in the space of Fig 162 black-in or color-in all the cells that are occupied by the number "2". In the space of Fig 163 color-in all the cells that are occupied by the number "7".

When you have located the six cells of "2" you can begin to connect every cell or point to every other cell in the space of Fig 164.

Likewise, in the space of Fig 165 draw in neatly all the connective lines of energy for the points of "7".

What observation can you make about the "X" of 2 & 7 ?

(Answer : The "X" of 2 is identical to the "X" of 7 except they differ by a 90° phase shift of consciousness. Also, you will have noticed that the "X" of 1 is a 90° turn of the "X" of 8. From this, you can expect that if 1 + 8 = 9 and 2+7 = 9, then the forthcoming diagrams of "X" of 3 will be similar to "X" of 6 & "X" of 4 " " " " "X" of 5).

On your right is Fig 166 which is the "X" of 2 (Fig 164) at 6 × 30°, or it can be termed : The "X" of 2+7 at 3 × 30°.

Fig 166

Notice how Fig 166 is very similar to the geometric design on the previous page. It is a detail from a Koran illumination from Egypt in the C15th. Holy Art lives ∞

To create the diamond of 81 Dots (9 x 9), you are required to join the edges, in pencil, criss-crossing parallel lines, then mark in black the 81 dots to begin superimposing the original pattern upon itself at 45°. Do this superimposition only after you have drawn the "X" of 2+7 at 0° in the given square of 81 Dots.

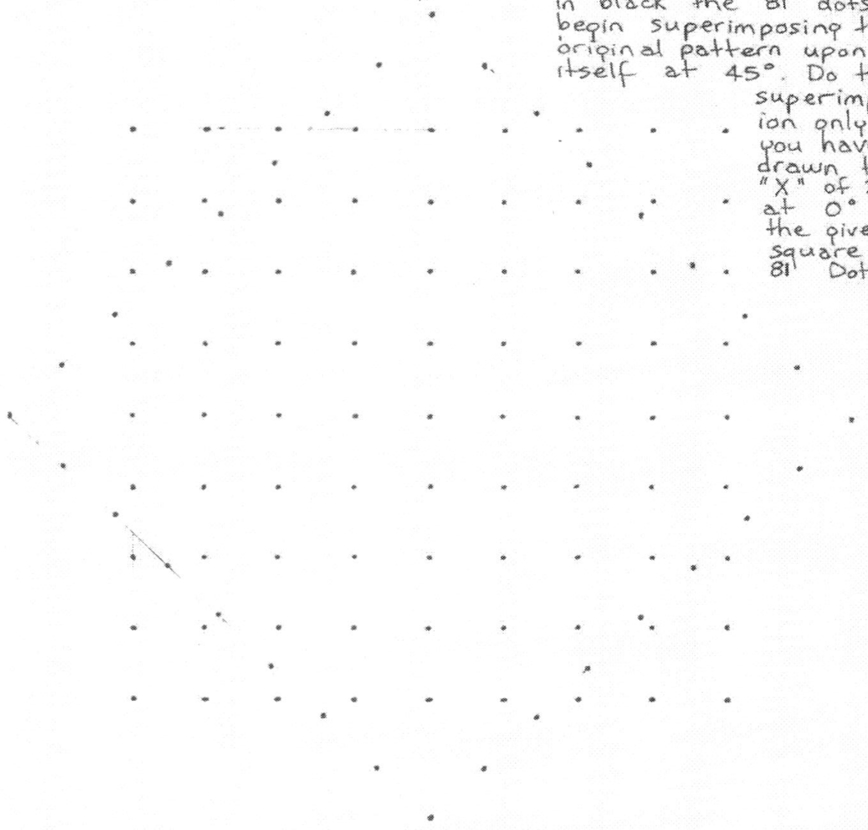

Fig 167

n the space of Fig 167 superimpose the yantras of Figs 164 & 165. This will be termed The "X" of 2+7 at 0°. This will result in an X-shaped yantra. Then observing the other lattice of 9x9 dots which is the diamond over the main square, you will be required to repeat what you have just drawn but at an angle of 45°. This superimposition will give a final design known as The "X" of 2+7 at 0 + 45°.

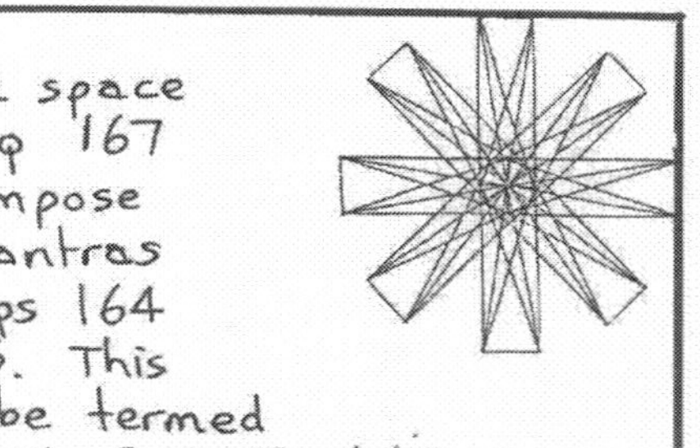

"The definition of the CROSS is Man-Woman standing in the attitude of DEVOTION standing in space with arms upraised in blessing sending out his-her gifts to humanity, pouring all his-her gifts forth freely in every directionnoitcerib".

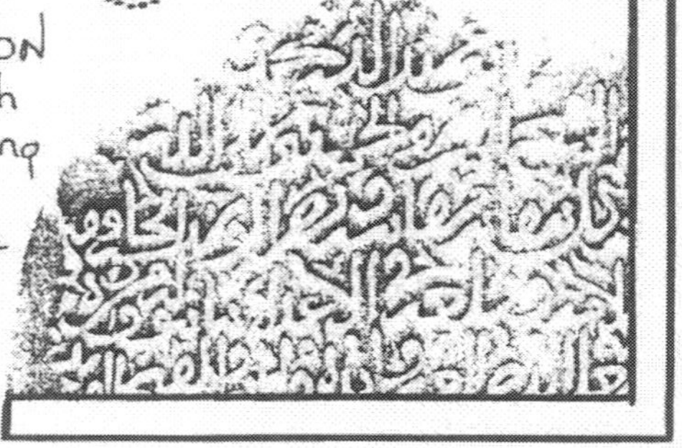

INTRODUCING the MULTIPLICATION TABLE & its Reduction Grid of THREE (3) and SIX (6).

fig 147

1	2	3	4	5	6	7	8	9
2	4	6	8	1	3	5	7	9
3	6	9	3	6	9	3	6	9
4	8	3	7	2	6	1	5	9
5	1	6	2	7	3	8	4	9
6	3	9	6	3	9	6	3	9
7	5	3	1	8	6	4	2	9
8	7	6	5	4	3	2	1	9
9	9	9	9	9	9	9	9	9

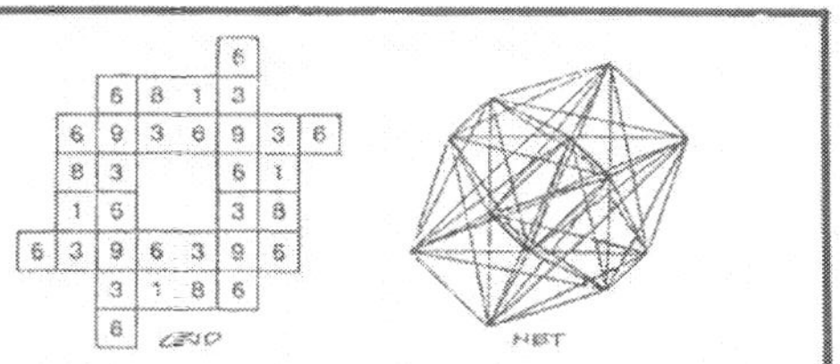

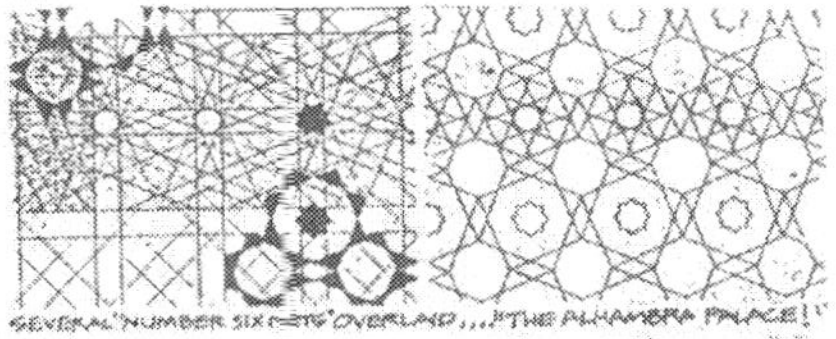
SEVERAL "NUMBER SIX NETS" OVERLAID ... "THE ALHAMBRA PALACE!"

LIFESTYLE DESIGN

... "He shows us a piece of paper ruled off into small squares. In each square is a number.

"This is a unique kind of magic square," he explains, and it is the key to my work of designing.

I have the task of designing this archway to transmit certain information & to give a certain feeling," he says with a gleam in his dark eyes. "Let us call the information & the feeling 'six'."

We watch closely as he abstracts from the magic square all the rows & columns which contain a "six". He comes up with a kind of number grid, which represents "six" & nothing else.

Then he takes a fine pen & skillfully joins all the points of the squares of the "number six grid," forming a kind of net.

He overlays several "number six nets" at different angles to one another, simplifies the design and comes up with this.

PROJECT

Soft

om:

Light

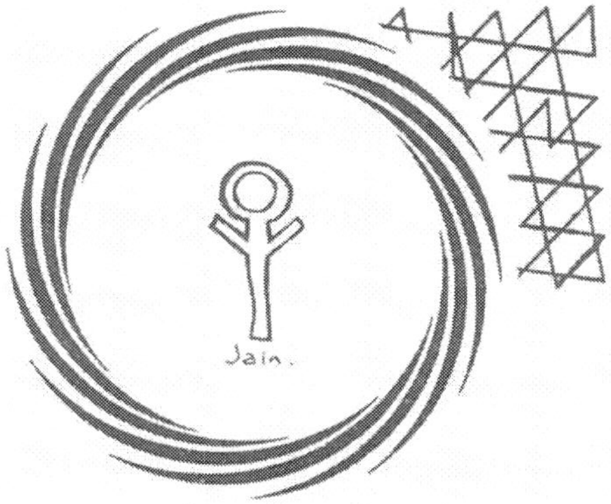

diagram of ornament of the octagonal Dome of the Alhambra.

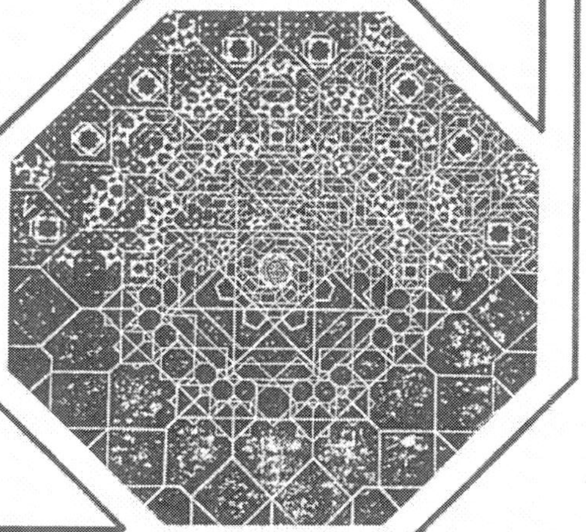

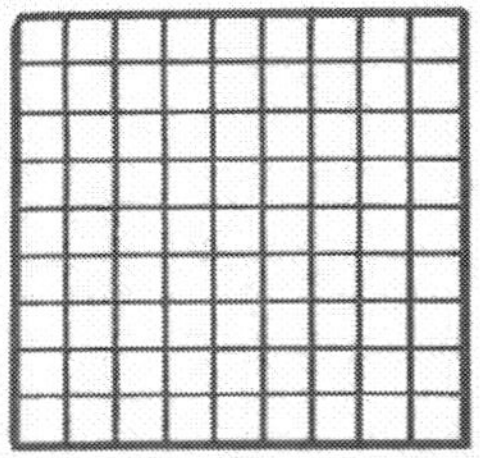

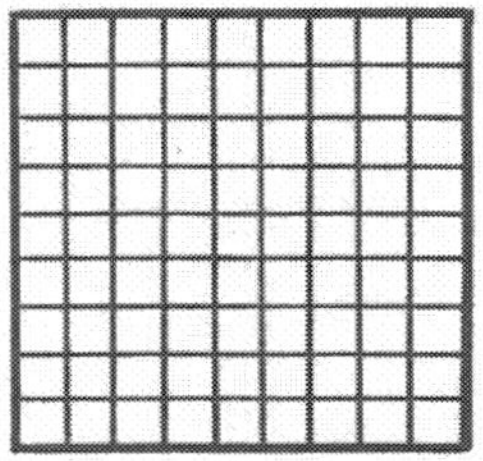

Fig 168

Fig 169

In the space of Fig 168 black-in or color-in all the cells occupied by the number "3" in The Multiplication Table / Reduction Grid of Fig 147 also shown on the previous page.

Similarly, in the space of Fig 169 color-in all the cells occupied by the number "6"

In the space of Fig 170 draw the "net" or "yantra" of Fig 168 (the Reduction Grid of 3) by joining each point to every other point.

Fig 171a shows the combined yantras of "3 + 6".

You will draw Fig 171a in an enlarged form in the following space of fig 171.

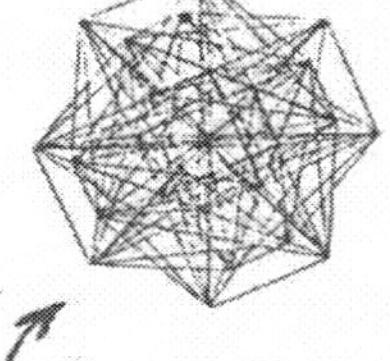

Fig 171a

Fig 170

an INTERVIEW WITH JAIN BY JAIN.

QUESTION : "What do you mean by : THE LOOM OF MAJIK SQUARES ?"

ANSWER : Firstly, let me define what a Loom is. A Loom is an instrument for weaving at least two different threads. The vertical thread (Warp) are the columns of the harmony squares that have equal sums, similar to the horizontal threads (weft or woof) or rows & of course my Loom has diagonal threads.

A tapestry is formed when the chaotic arrangement of numbers in the Loom Box are joined or sewn together in arithmetical counting order.

My Life's Purpose, the Heart of "F.R.E.E.D.O.M.S." (For Research Expressing Essential Data Of Majik Squares) is to codify & establish a functioning system of Majik Square Knowledge, Science & Art in Stone, Cyber Space & Text. The image on your left: Escher's spiral, symbolizes to me how Majik Squares are related to the health & Wholeness of our Globe, how a 2-Dimensional lattice/matrix/loom can pierce realities into 3-D, 4-D & → ∞

These Looms of Harmony Squares exist in & define a field of consciousness that transcends both time & the individual. Plato & Jung would have called this an : "ARCHETYPE". That is the ancient attraction for these Majik Squares, even though you have not studied them, the knowledge exists within your cellular memory. My Loom is for Children; even a 5 years young child can construct a M.Sq. of 3 at 0° + 90° Yantram. I object to the arcane mysterious occult sciences that nobody can fathom out; so my tapestry is very clear, simple & intelligent. Call this Weaving Art "Genetic", call it "Galactic", whatever word we use, I will continue this LifeTime to preach of its Timeless Simplicity. The actual weaving of numbers (which is The Dance of your Journey) to form M.Sq. Art constitutes a mythic voyage to the Galactic Heart Centre Within. This is how I view M. Sqs, that they provide entrances or portals to multiple realms of being & consciousness.

The Visual Multiplication - Table from its Reduction - Grid of 3+6.

Fig 171

INTRODUCING the MULTIPLICATION TABLE & the Reduction Grid of FOUR (4) and FIVE (5).

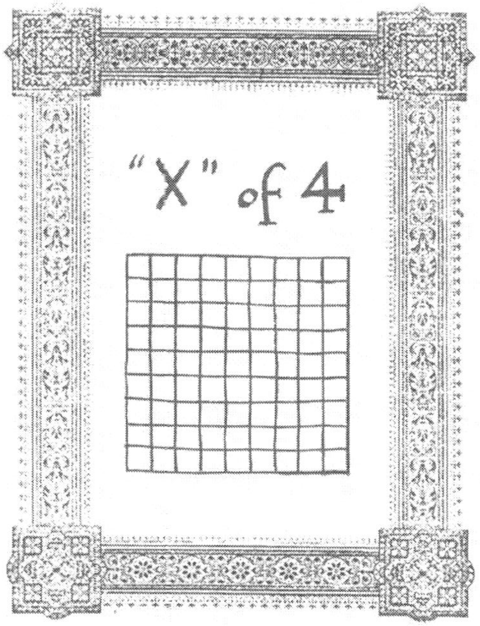

Fig 172

Similar to the "X" of 1 + 8, 2 + 7, 3+6, plot the pattern for all the points of "4" in the space of Fig 172. You will have to refer to the Reduction Grid of Fig 147 & locate all the cells occupied by the number "4".

In the space of Fig 173 color-in the pattern for all the points of "5".

"X" of 4 & 5

FUN WITH MATHEMATICS

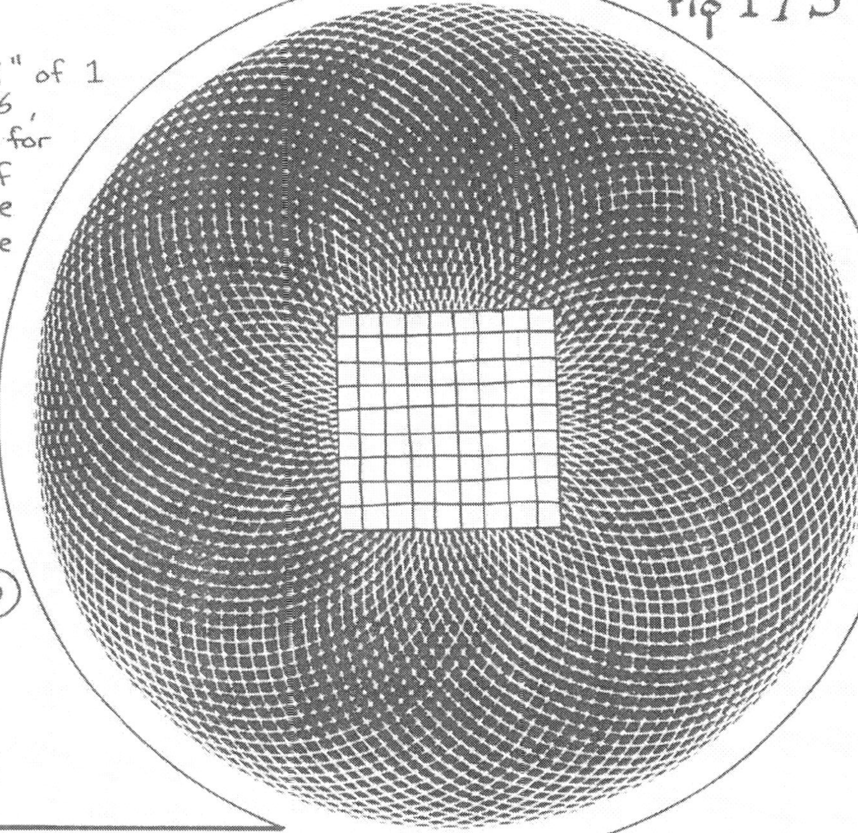

Fig 173

In the space of Fig 174 connect all the points of "4" (Fig 172).

In the space of Fig 175 connect all the points of "5" (fig 173) such that each point connects every other point.

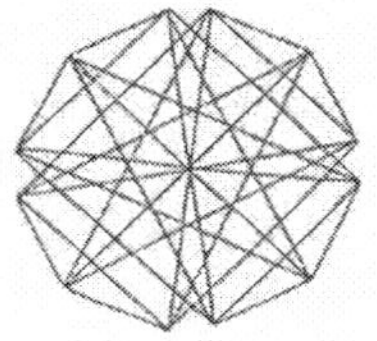

In the space of Fig 176 → superimpose fig 174 & 175 to form the yantram known as: The "X" of 4 + 5 at 0°.

Fig 174

Fig 175

Fig 176 →

"X" of 4 + 5 at 0° Yantram muralized on a brick wall → (Fig 176, also shown below).

These Multiplication Yantras represent Archetypes of Growth; that's what "to Multiply" means: to grow. Crystals are living entities that are in a constant state of growth. Thus these yantras are the INFRASTRUCTURE of all living things,

this universal CRYSTALLOVE,

THIS WATERFALL

OF DIVINE WISDOM

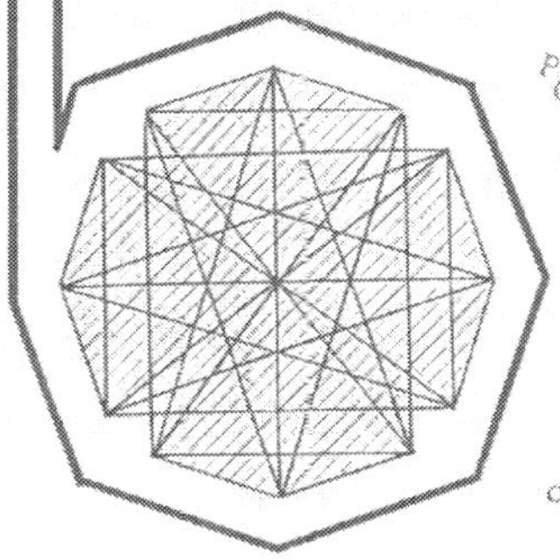

POURING

in2

our CROWN,

These MAGIK SQUARES are adapted from the many wise traditions of the earth and its cultures. They celebrate the ever-renewing truths of independence, wholeness and unity: in full colour to mirror the natural transparencies of the mind. They add a subtle touch of colour to stimulate the senses.

Fig 176

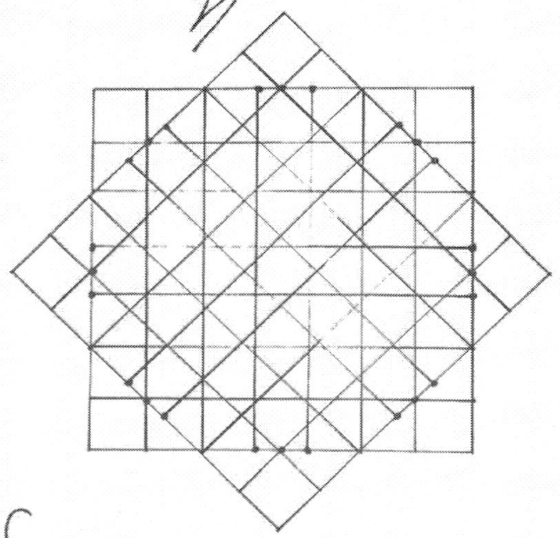

Fig 176 above is similar to fig 160a which shows the geometry necessary to construct a yantram at 45° to its 0° position. It can also be referred to as 'constructing a diamond over its square'. Each square has 7x7 cells to give 8x8 points for the Lattice of dots. (The 9x9 Lattice of dots has been simplified to 8x8 by ruling out the column & row of nines in Fig 147). In the space of Fig 177 draw the "X" of 4+5 at 0° + 45°. It means that what you drew in Fig 176 is to be superimposed by its rotation of 45°. The dots necessary for this Yantram have been given in both figs 174 & 175, & have been highlighted on this page.

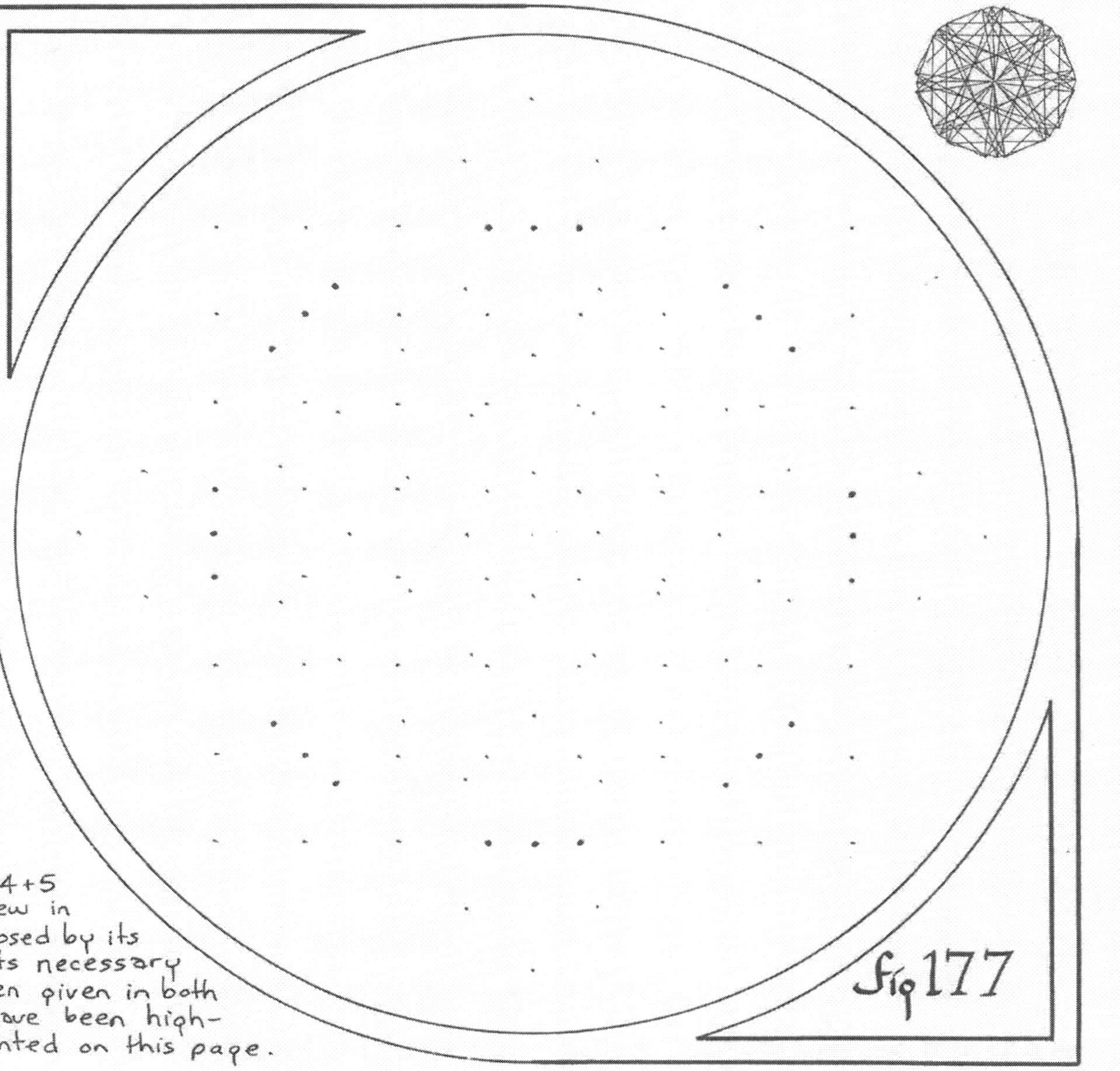

Fig 177

INTRODUCING the MULTIPLICATION TABLE & its Reduction Grid of → NINE (9)

Referring to the Reduction Grid of Fig 147 locate all the 21 cells that are occupied by the number "9". Then color-in these cells in the space of Fig 178.

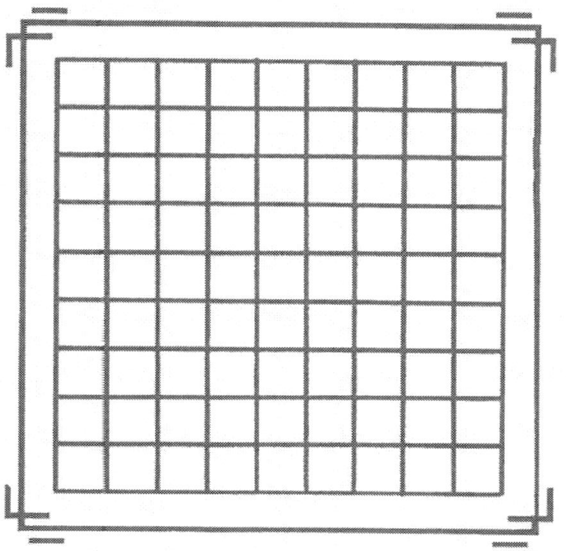

Fig 178

KEEP ON MEDITATING
MY FRIEND
NOT TOO MUCH
NOT TOO LITTLE
KEEP ON FLOWING
MY FRIEND
ENDLESS
ENDLESSLY BE
COMING
endlessly becoming
NOWHERE = NOW HERE
endlessbecoming
endlessbecomingendlessbecom
ingendlessbecomingendlessb
ecomingendlessbecomingendle
ssbe. BE WHAT YOU ARE:
LOVE. JUST LOVE.
Love from HAHA-AHA. 4.3.90 Ko Samui

"X" of 9

Referring to the Reduction Grid of Fig 147 & all the Points of 9 of Fig 178 carefully construct the "X" of 9 Yantram. This design will require many more lines drawn than any previous yantram so ensure that your pen or pencil is very fine / sharp. In the holy space of Fig 179 nurture meditative thoughts.
Color-in & Enjoy.

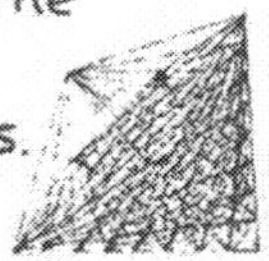

Fig 179 →

Fig 180 →

The purpose of this book is to add a creative element or touch to these designs that you have just drawn.

By taking Fig 179 & tiling it four times into a larger square the result is Fig 180 shown here →. I call this " The Four Angels of Nine ".

e
en
end
endl
endle
endles
endless
endlessl
endlessly
Becoming.

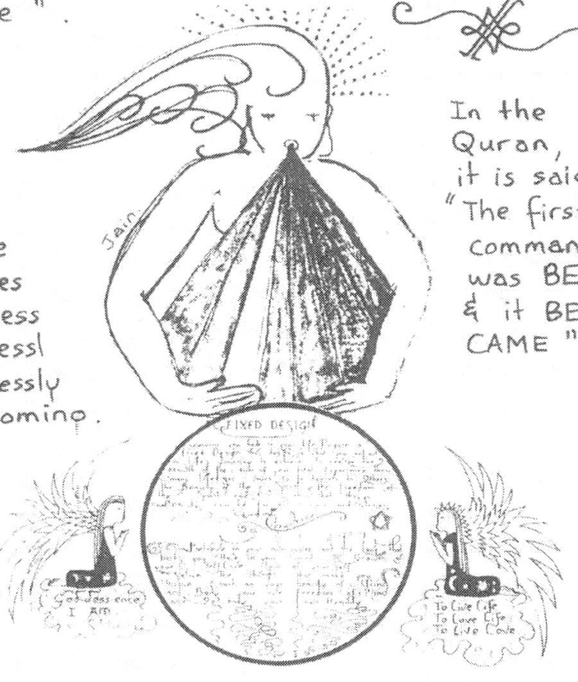

In the Quran, it is said: " The first command was BE & it BECAME ".

Border in temple form. Academicism, Munich 1839

CHAPTER ZERO:

SOLUTIONS TO THE MAGIC SQUARES

PAGE 197

SOLUTIONS : CHAPTER ELEVEN

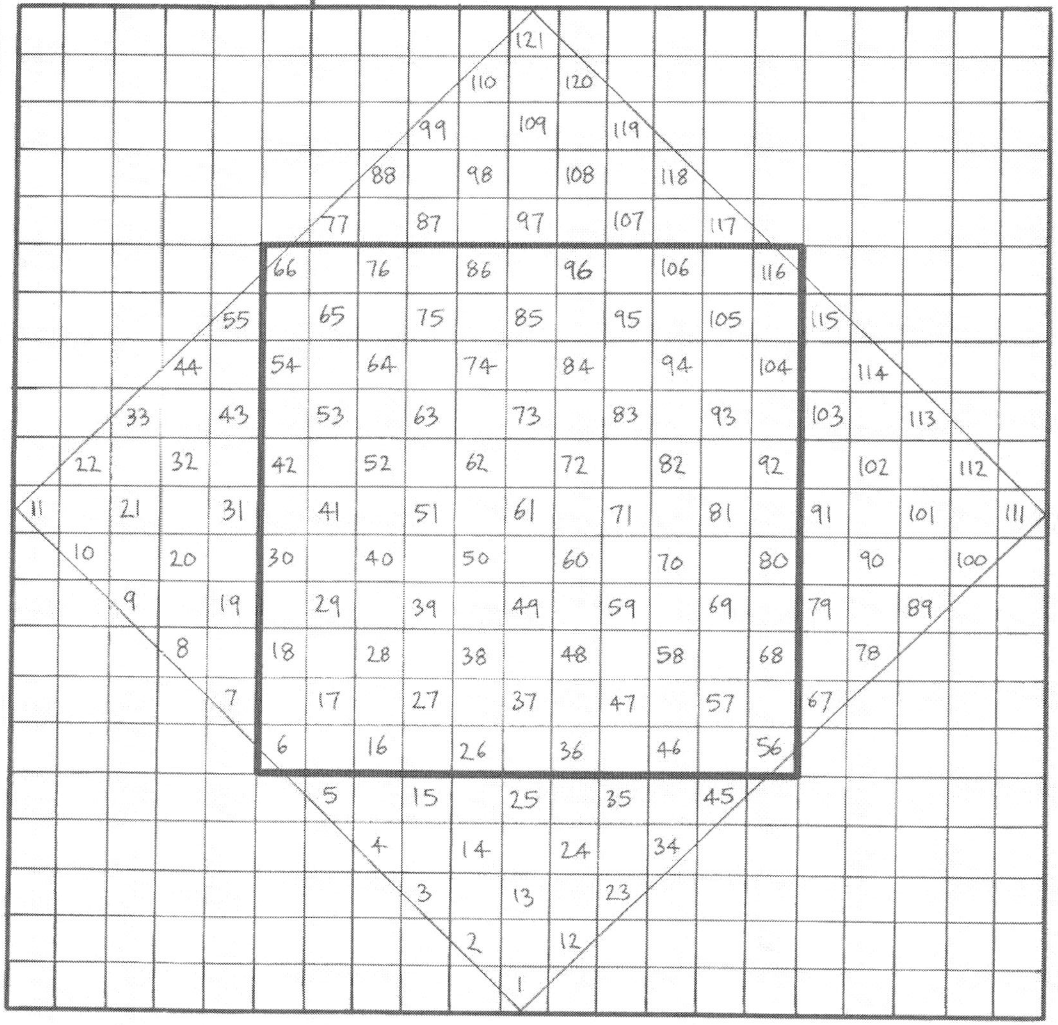

CONSTRUCTION OF THE ESSENE MAJIK SQUARE OF ELEVEN.

As indicated in Chap. 9 (Fig 48) via the Method of Transposition, this diagram Fig 111a shows a Frame Size of 21 x 21 to contain the Natural Square of 11^2 (ie. 121 numbers). The Central Square in bold (11 x 11) will be the Majik Square when all numbers outside of this square are transposed 11 cells towards the centre.

1x1	=	1
11x11	=	121
111x111	=	12321
1111x1111	=	1234321
11111x11111	=	123454321
111111x111111	=	12345654321
1111111x1111111	=	1234567654321
11111111x11111111	=	123456787654321
111111111x111111111	=	12345678987654321

Shown here are 3 examples of LUSUS NUMERORUM (Latin for Number Jests) which involve some "Regular Peculiarities" of the Number 1 or 11. These magic Numbers serve to indicate the Inner & Timeless Beauty of the Universal Language of Mathematics.

$1 \times 8 + 1 = 9$
$12 \times 8 + 2 = 98$
$123 \times 8 + 3 = 987$
$1234 \times 8 + 4 = 9876$
$12345 \times 8 + 5 = 98765$
$123456 \times 8 + 6 = 987654$
$1234567 \times 8 + 7 = 9876543$
$12345678 \times 8 + 8 = 98765432$
$123456789 \times 8 + 9 = 987654321.$

$0 \times 9 + 1 = 1$
$1 \times 9 + 2 = 11$
$12 \times 9 + 3 = 111$
$123 \times 9 + 4 = 1111$
$1234 \times 9 + 5 = 11111$
$12345 \times 9 + 6 = 111111$
$123456 \times 9 + 7 = 1111111$
$1234567 \times 9 + 8 = 11111111$
$12345678 \times 9 + 9 = 111111111$
$123456789 \times 9 + 10 = 1111111111.$

fig 111a

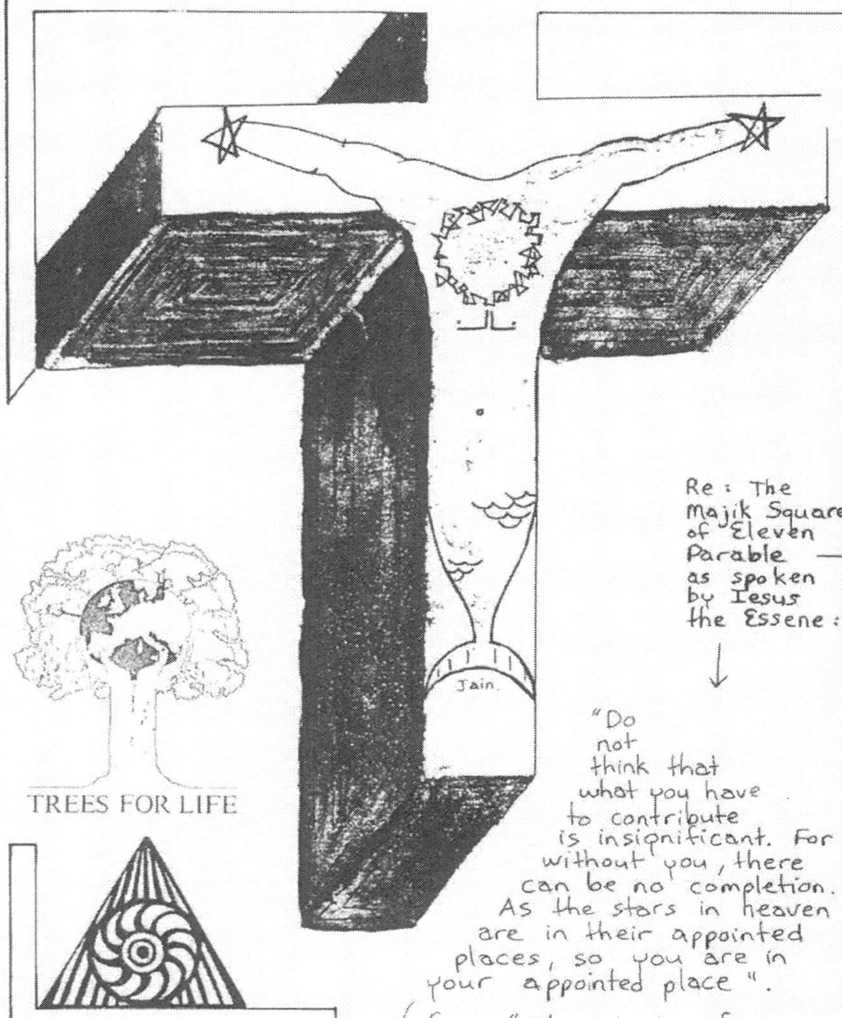

Re: The Majik Square of Eleven Parable → as spoken by Iesus the Essene:

"Do not think that what you have to contribute is insignificant. For without you, there can be no completion. As the stars in heaven are in their appointed places, so you are in your appointed place".

(from "The Words of CLEARWATER").

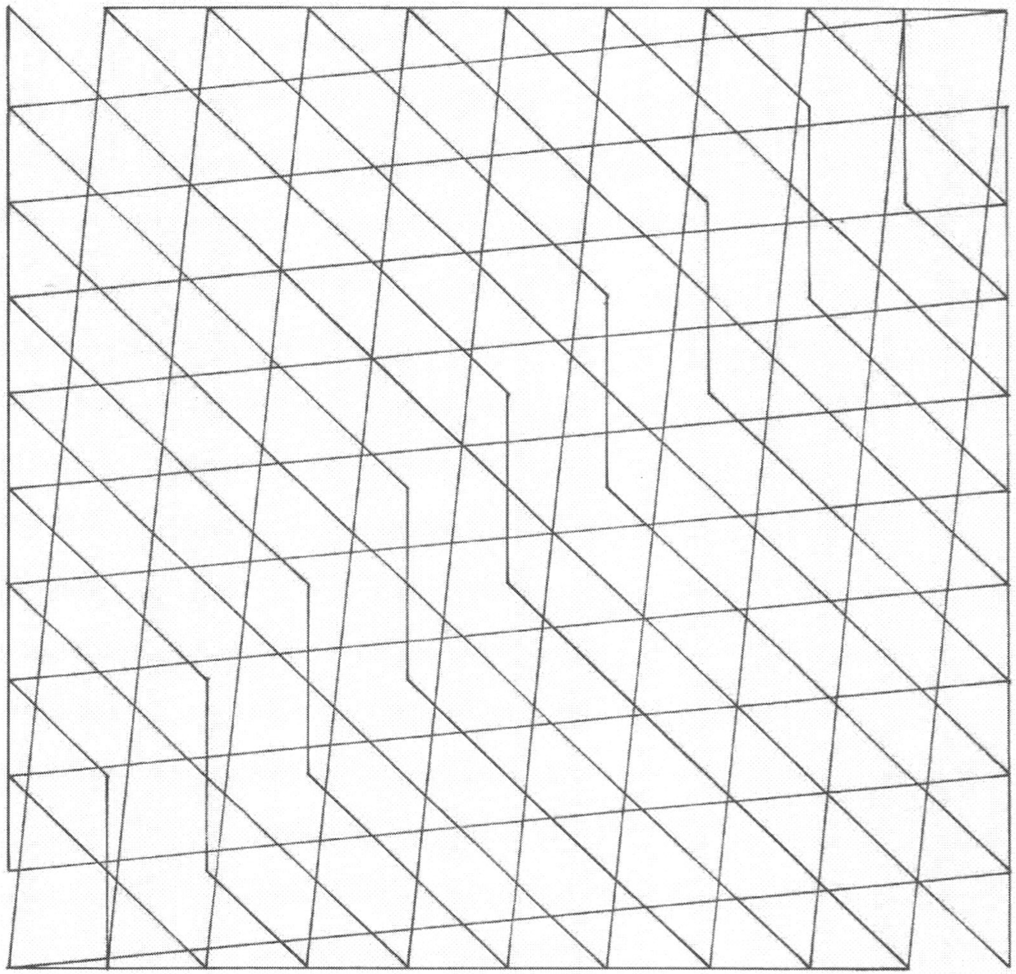

Fig 111

Fig 112

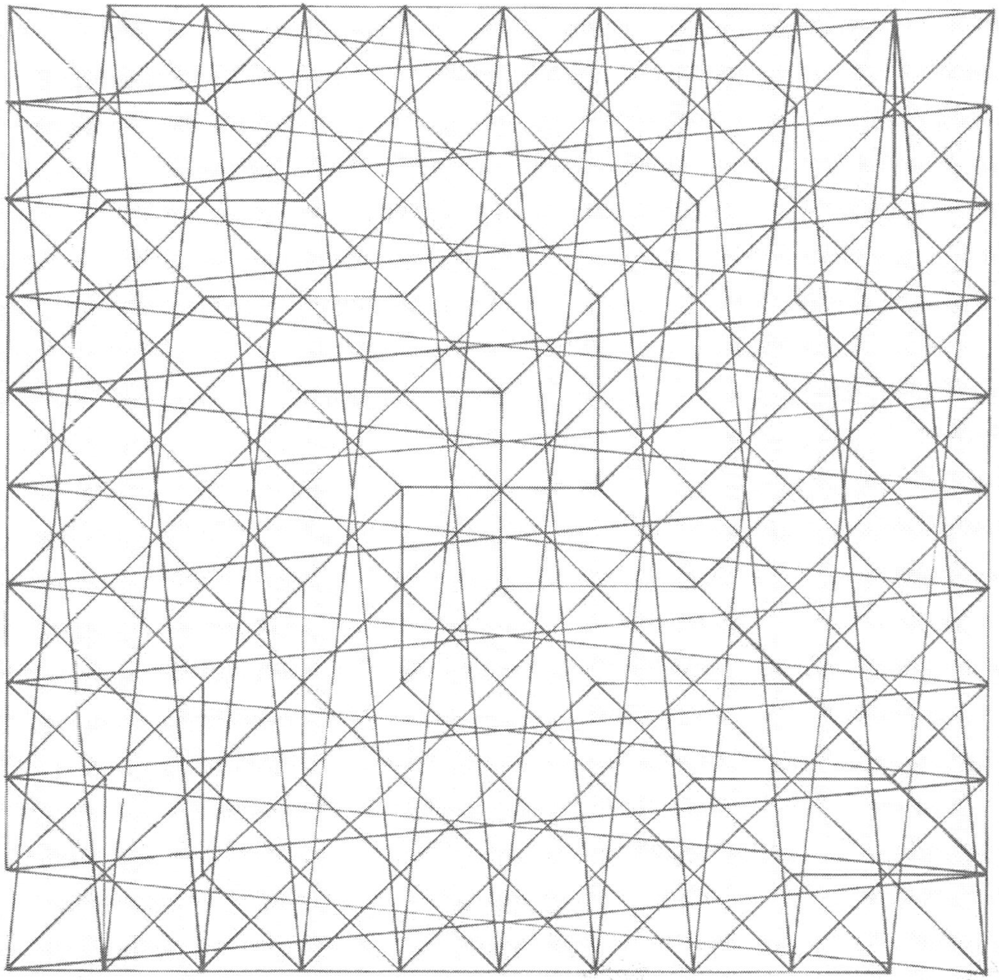

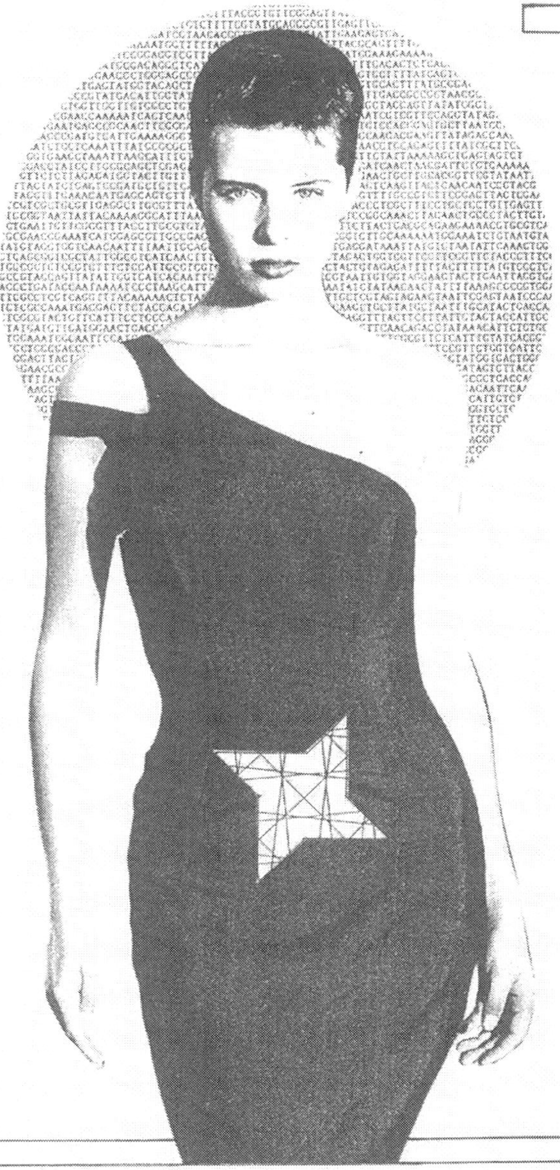

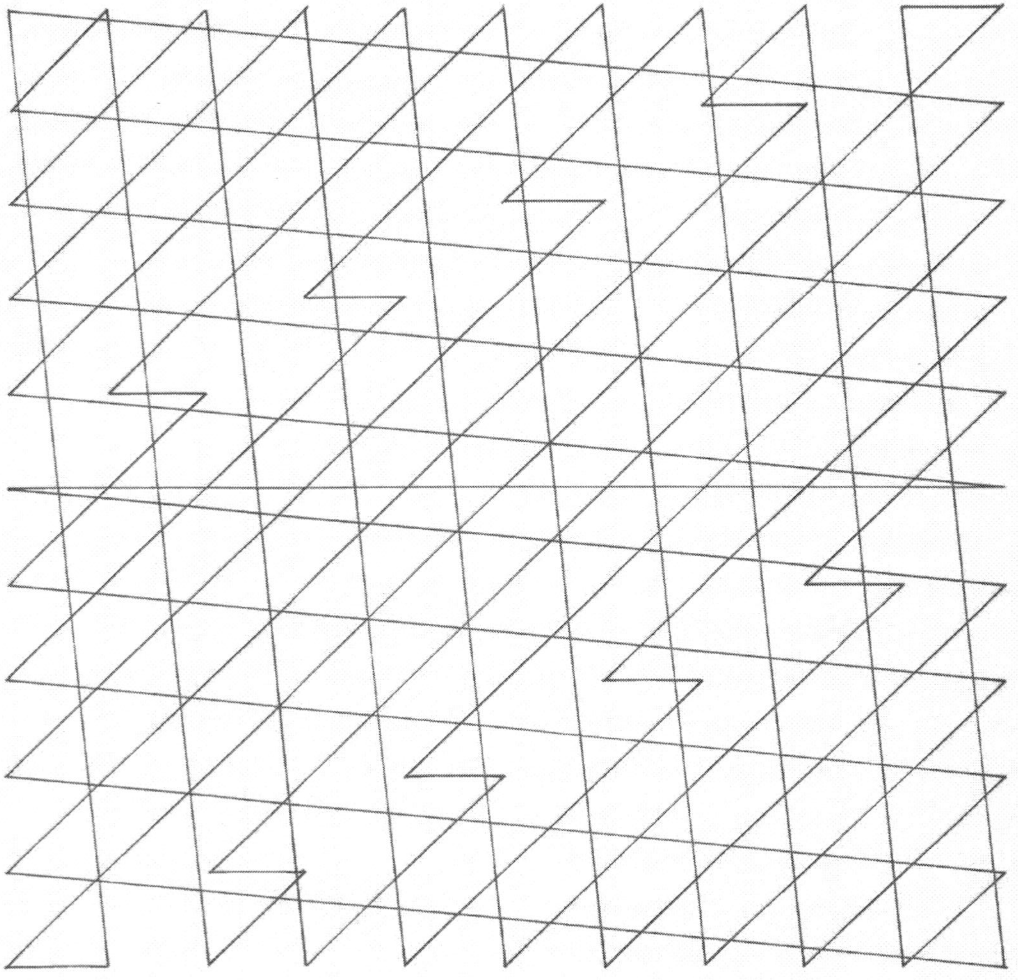

Fig 117

Glossary: "NUMINOUS"
= "a feeling of awe & wonderment characteristic of Man's / Woman's communion with God-Goddess-Goddessence".

at One with the Stream

Morning Meditation :)

I lay on the smooth rocks & my left hand, facing up, found its way into the cold flowing stream. It was a refreshing numinous feeling... As I then remembered that Energy flows into Human Temple via the Left-Hand Channel ... & out thru the right, my right-hand, facing down, resting 'pon my Heart, blissfull Healing Circuit, embracing Heaven & Earth, transforming

Six-point Star

THE SIX-POINT STAR
Using the Magic Number 26, there are 80 possible solutions. This is one.

To find an optimal total for a Magic Star's Magic Number, add all the numbers to be distributed, double that, and divide by the number of points of the Star.

fig 121

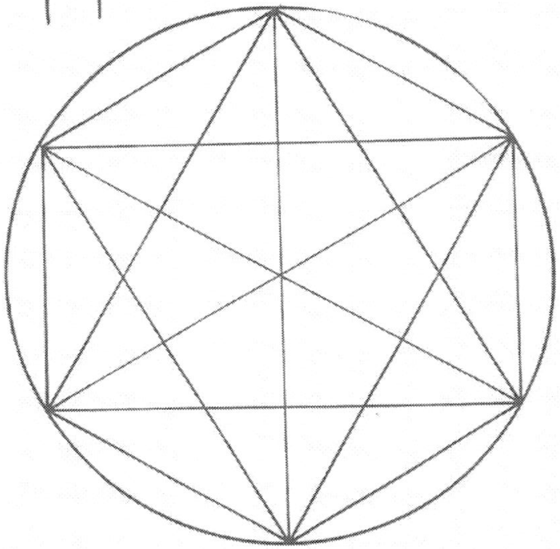

fig 120

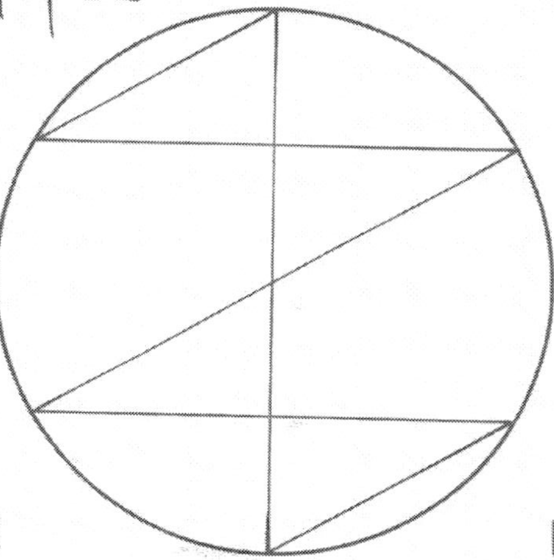

Magic Hexagon

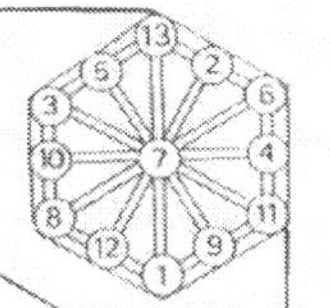

MIND BENDERS

Magic Cross

	12	9	
2	7	6	11
5	3	10	8
	4	1	

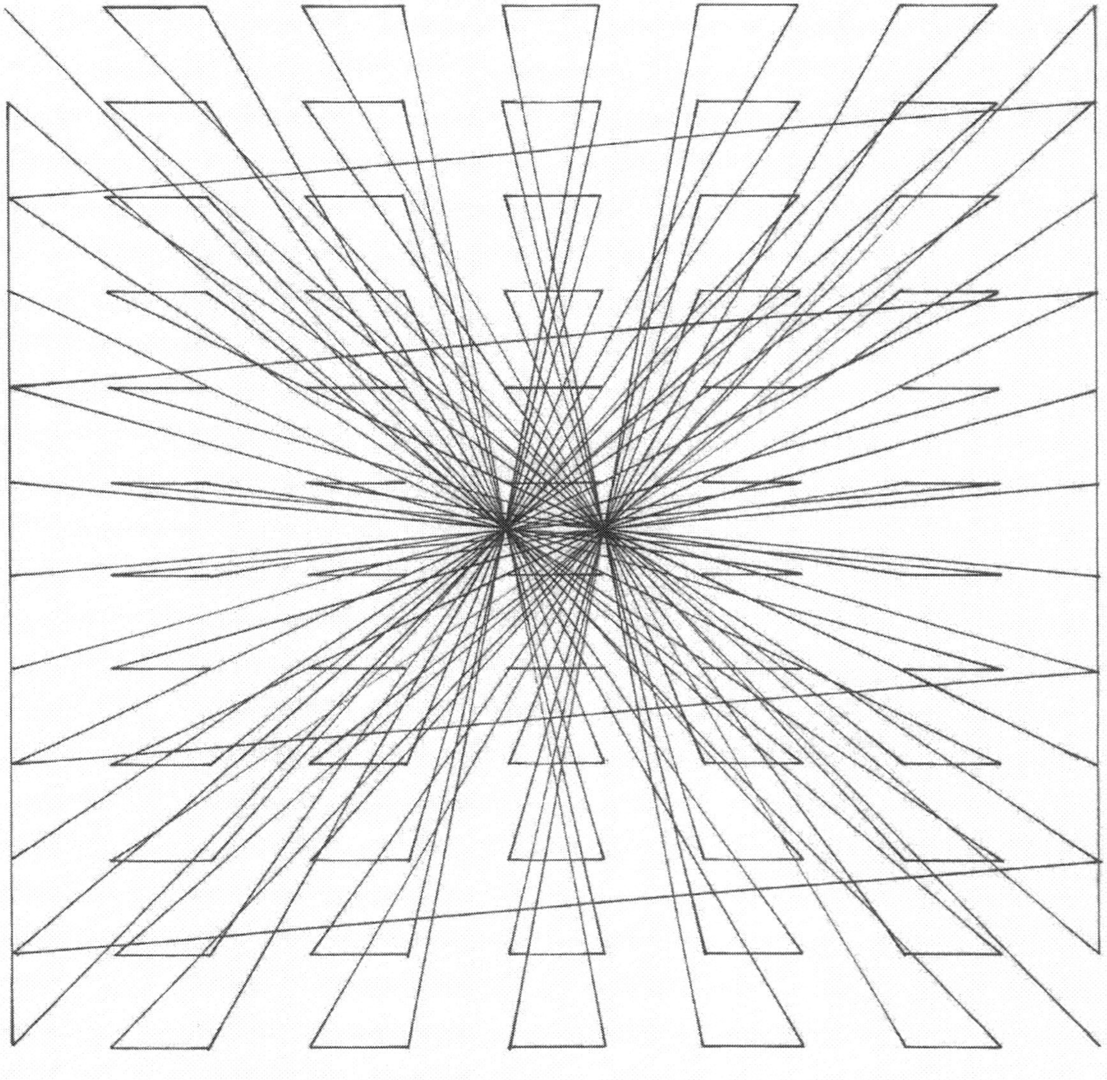

Fig 124

Forget all the rules
you ever learned about
graphic design.

Including the ones in
this book.

Forget how good design
is *supposed* to look.
What you think is good
design, is what other desig

FIG 126

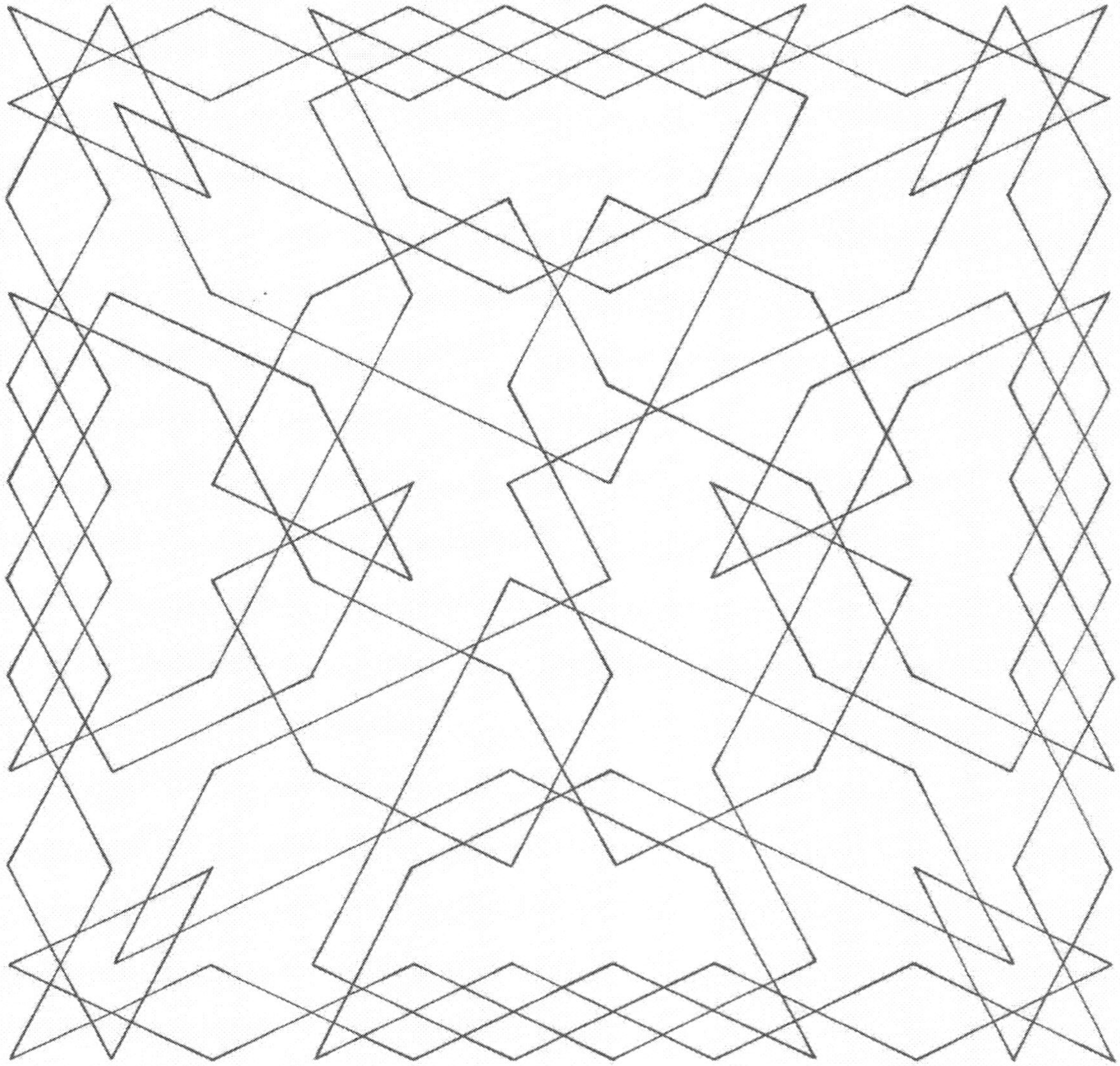

FIG 128

▲Even in the darkest period of mathematical learning, some people kept the science alive. Gerbert of Aquitaine (953–1003), whobecame the first French pope, Sylvester II, was a mathematician. He collected the surviving books of Latin and Greek authors to preserve their ideas. To help others do long and difficult calculations, he built globes of the earth and the heavens and popularized the use of the abacus.

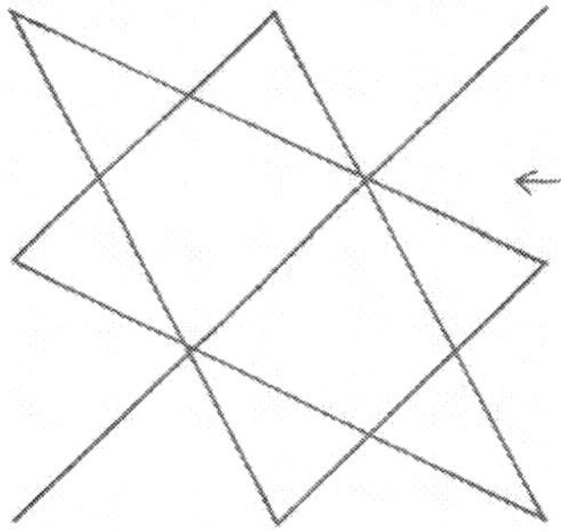

This Seal of Solomon is close to the Golden Mean Triangular shape. When tessellated the array is similar to the SILICON crystal field well known now as Atomic Art/ Structure!

fig 132

(NB: The Solution for Fig 130 is identical to both Figs 132 & 133).

fig 133

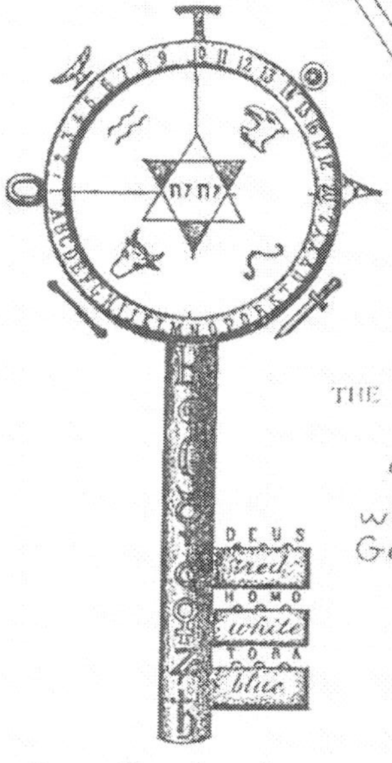

Absolute Key to Occult Science.

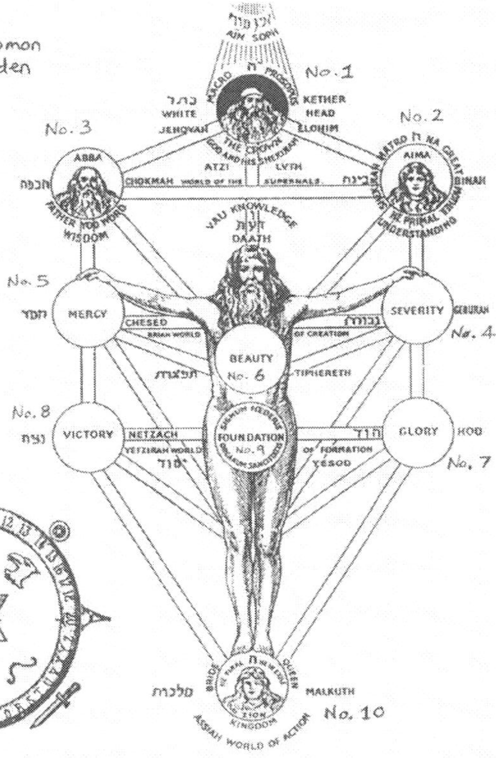

THE SACRED TREE OF THE SEPHIROTH

"A Great Truth is a Truth whose Opposite is also a Great Truth".
Niels Bohr
(Atomic Scientist).

"We must continually count on the appearance of new facts, the inclusion of which within the compass of our earlier experience may require a revision of our fundamental concepts"
Niels Bohr

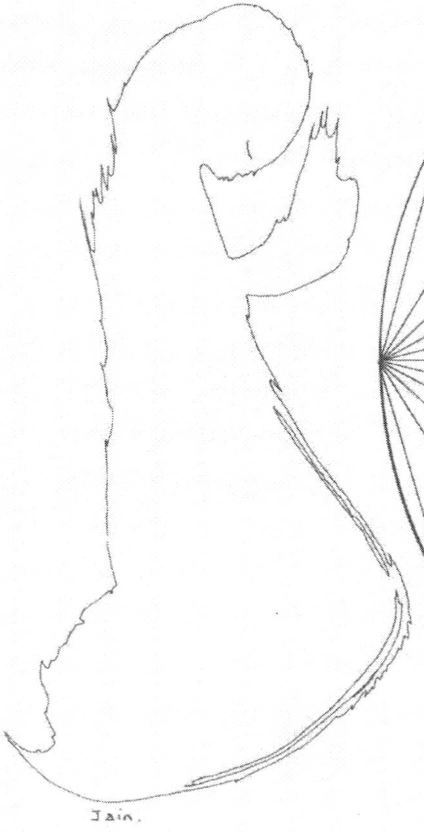

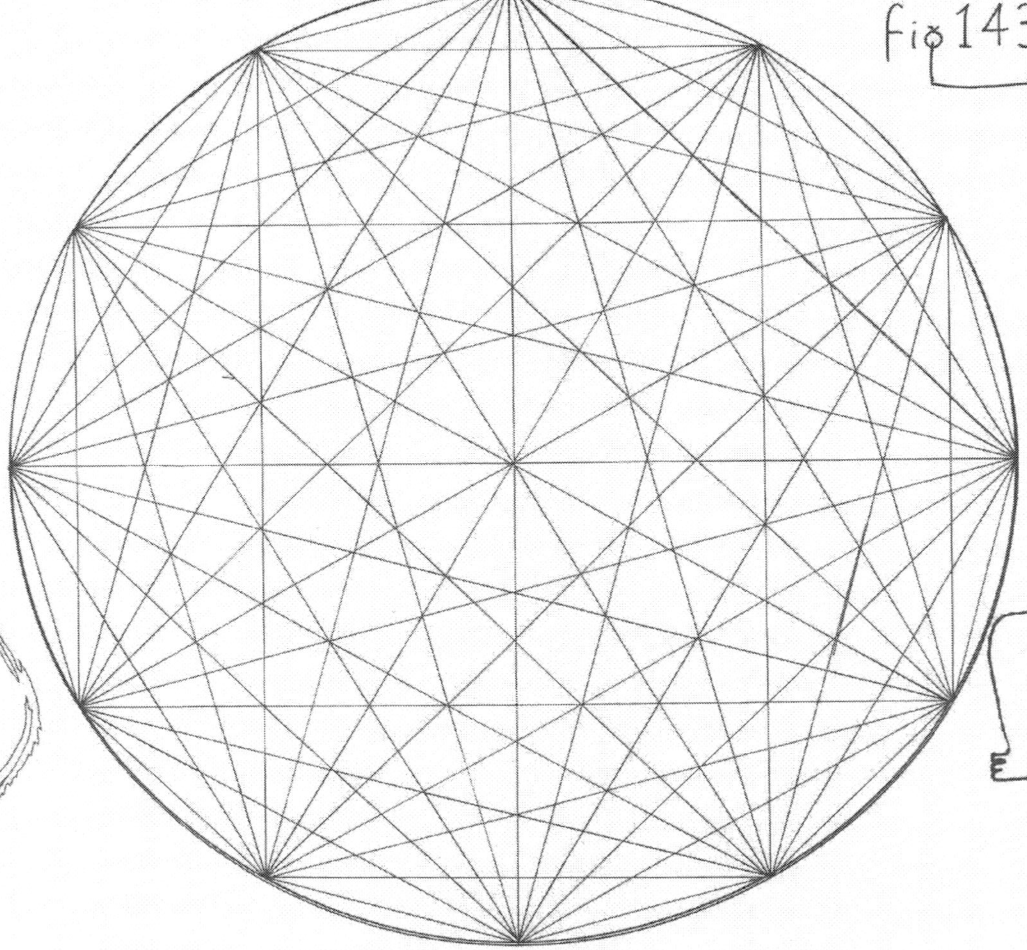

fig 136

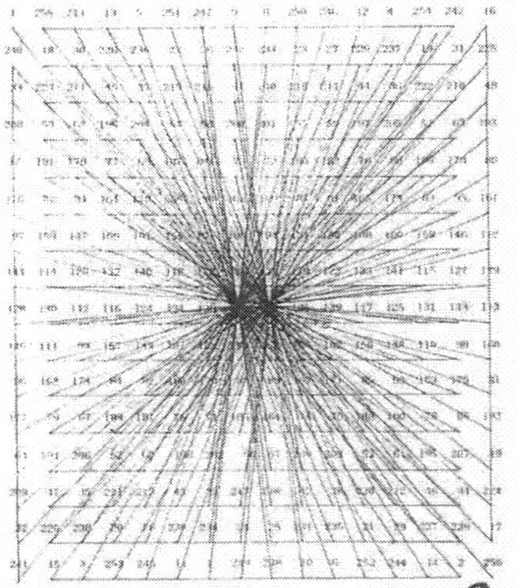

fig 143

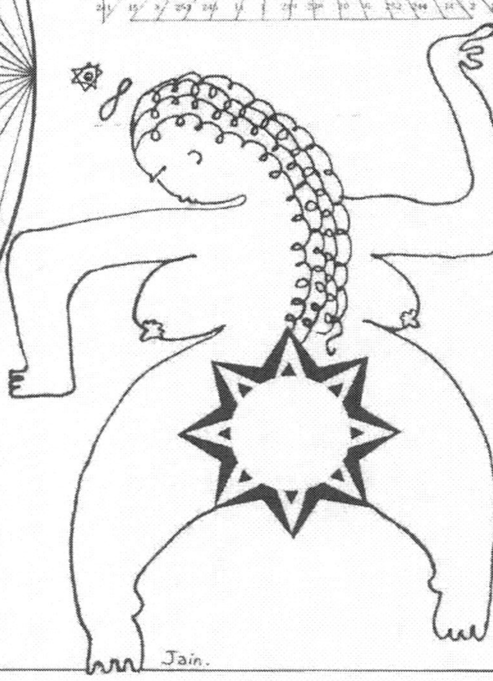

FIG 140

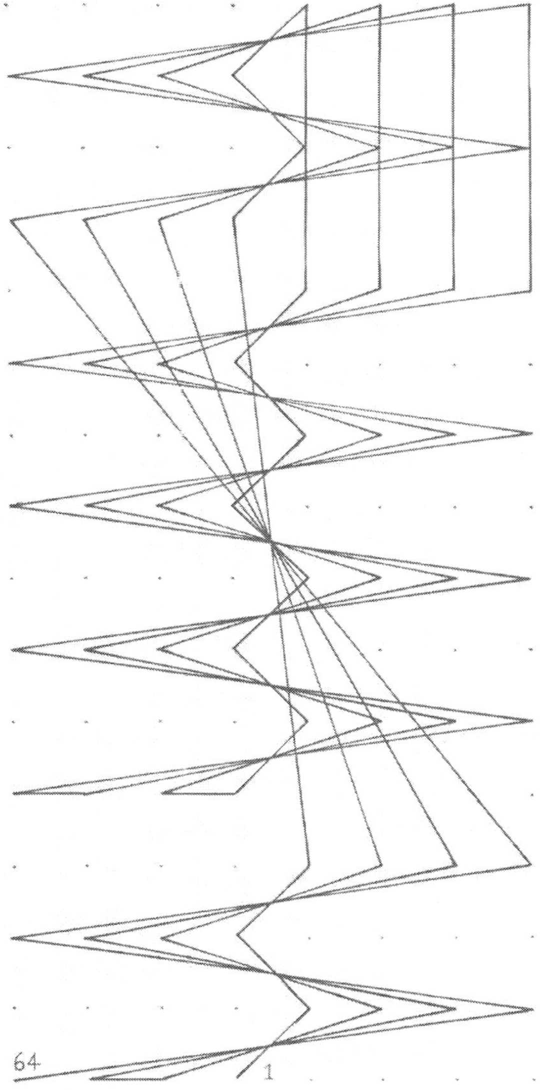

FIG 141

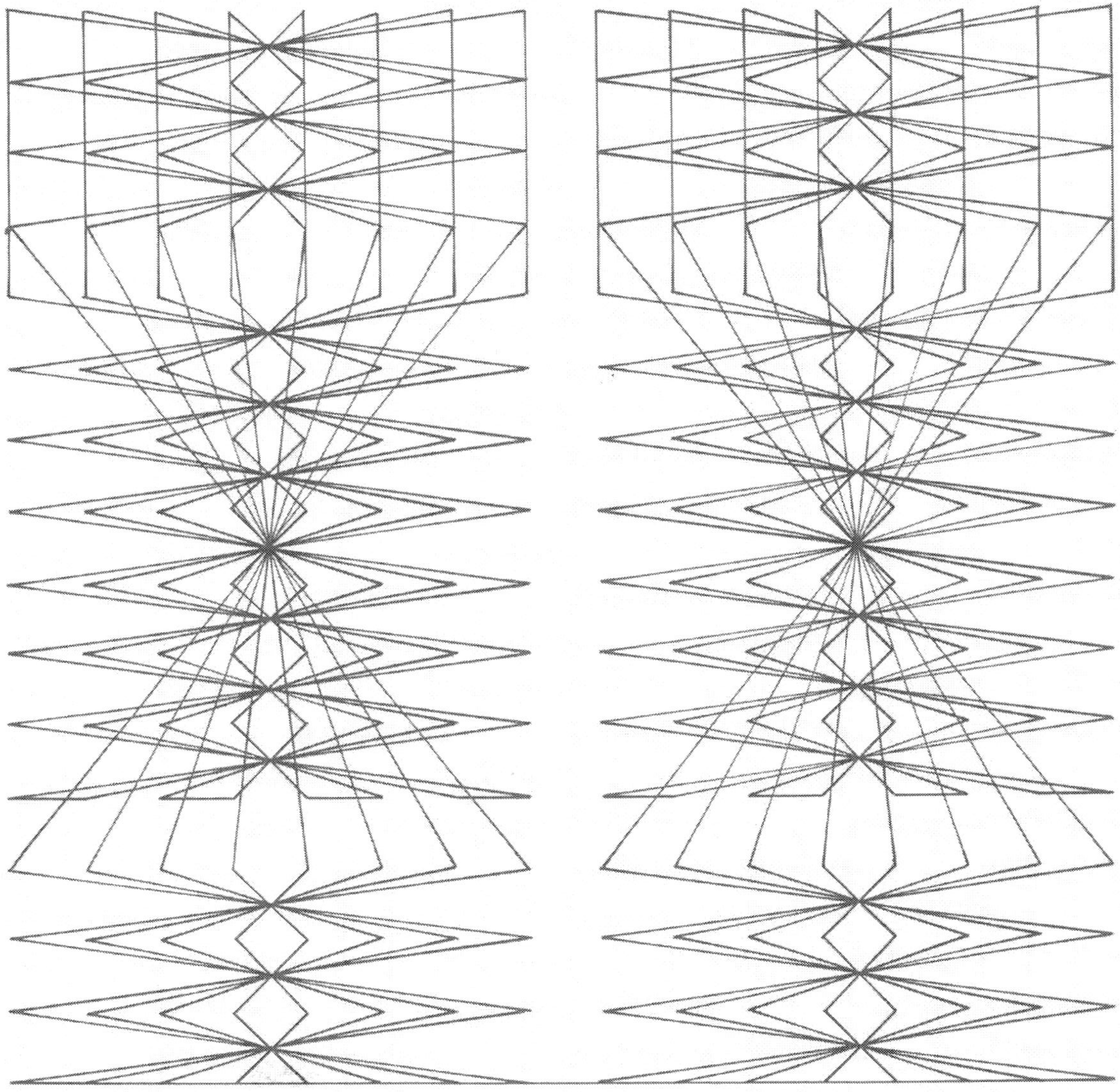

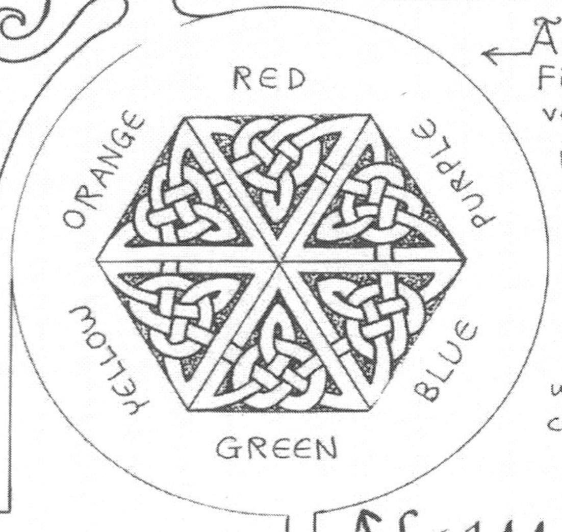

Fig 144

ANSWER to Fig 144: YES. Shown very simply, the complementary opposite color of Red is Green, the opposite of Orange is Blue, & the opposite of Yellow is Purple. (Red, Yellow & Blue are the 3 Primary Colors, that when mixed, form the secondary colors of Orange, Green & Purple.

This Science of Color Opposites explains why, after staring at a Red mandala for 3 mins. & you close your eyes, that the After-Image becomes a Green Mandala.

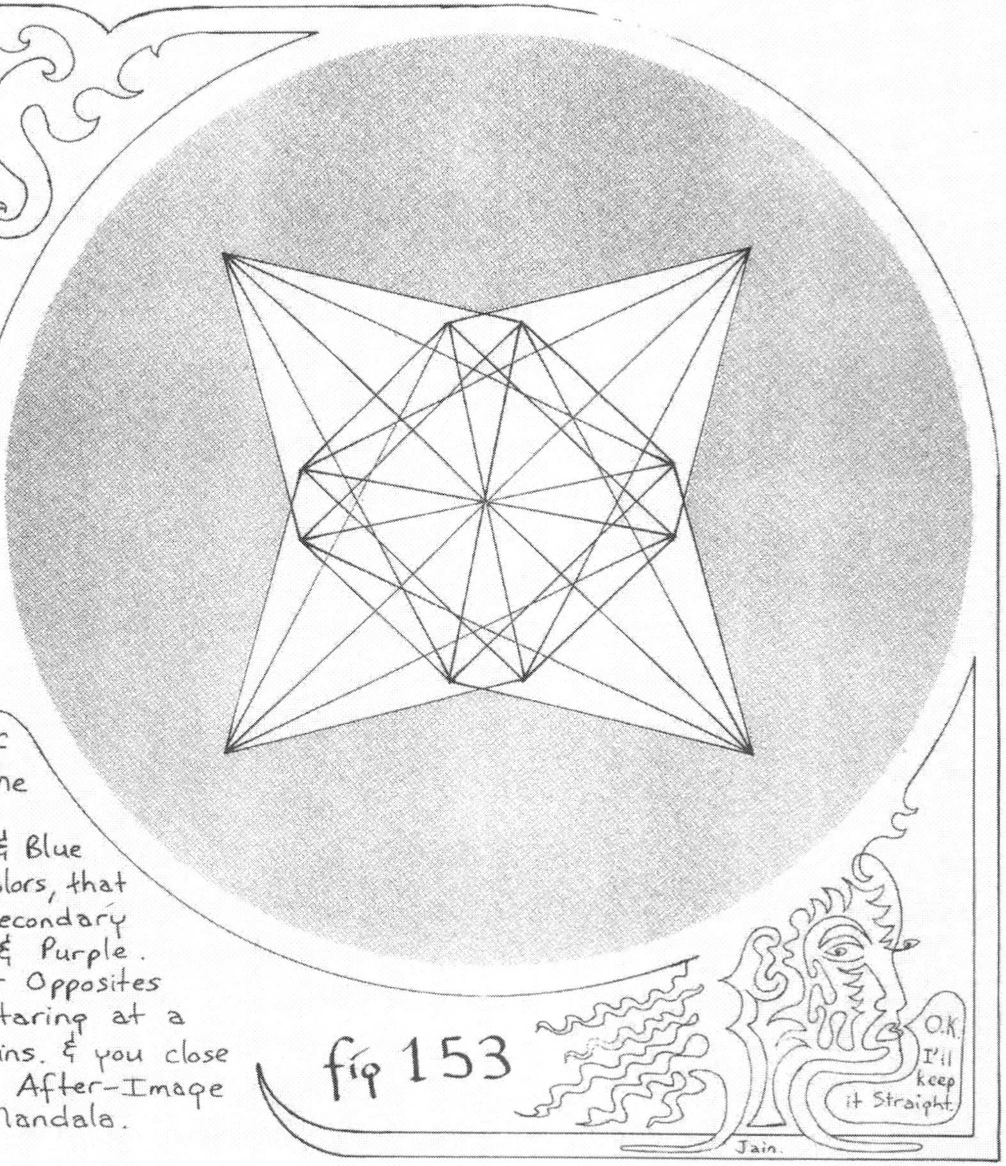

fig 153

I whole-heartedly dedicate this Yantram of Fig 160 ("x" Grid of 1 + 8 at 0° + 45°) to the AMBULANCE SERVICE of N.S.W. Australia, 'cos they saved my Life on the dark nite of 27·1·1984 in the Blu Mountains, where I was stabbed close to the heart, 6" deep, whilst painting a 3m x 3m mural, similar to this design. I have highlighted the central cross to illuminate this sacred symbol of Rescue Healing.

Opening the heart chakra

fig 160

Pattern for all The Points of 2
(the pattern for 7 is similar).

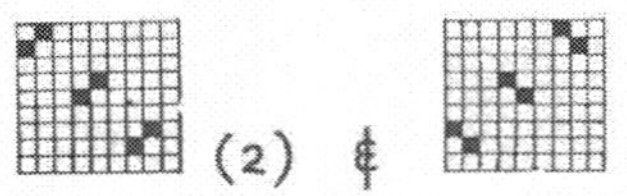

(2) & (7).

fig 162   fig 163

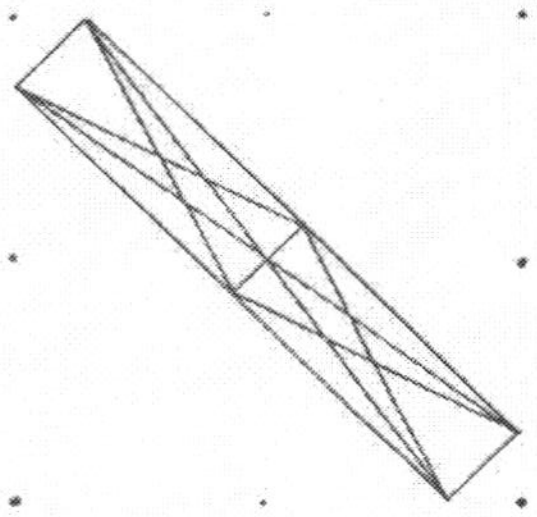

fig 164

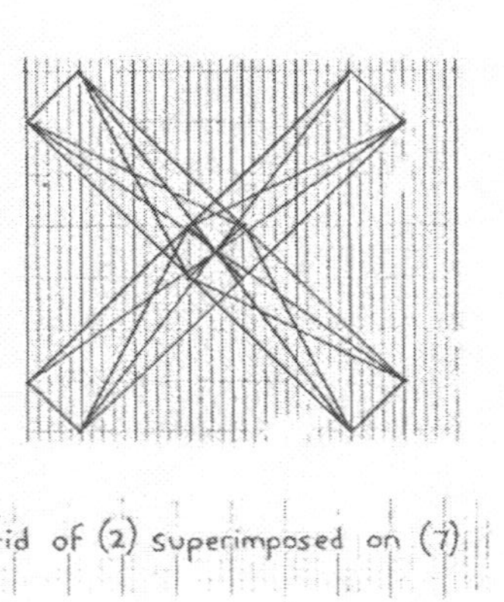

Grid of (2) superimposed on (7)

figs 164 + 165

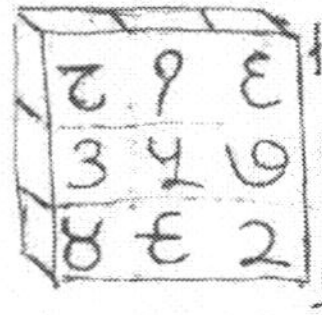

the unfolding of
Love,
the peace Within,
the Joy I know;
the World-Wide
Networking of the Majik
Square Institute, Industry
of Love - Geometry.

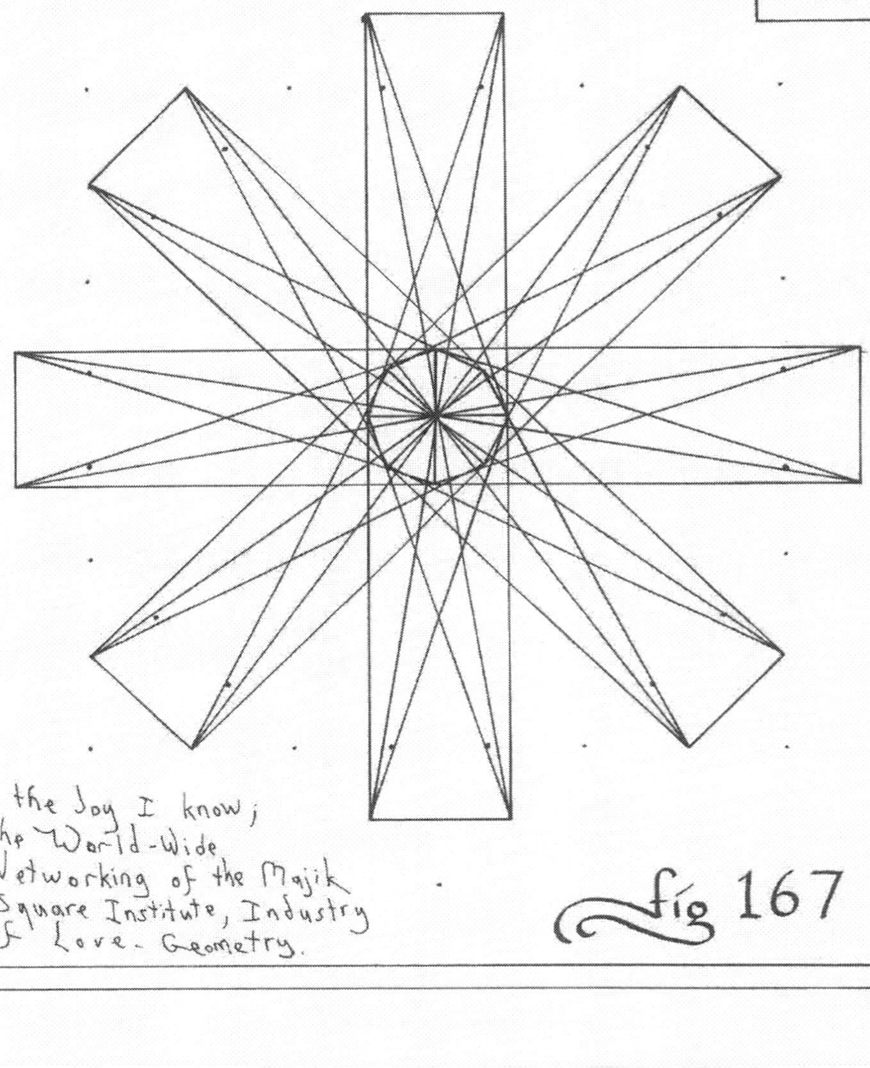

fig 167

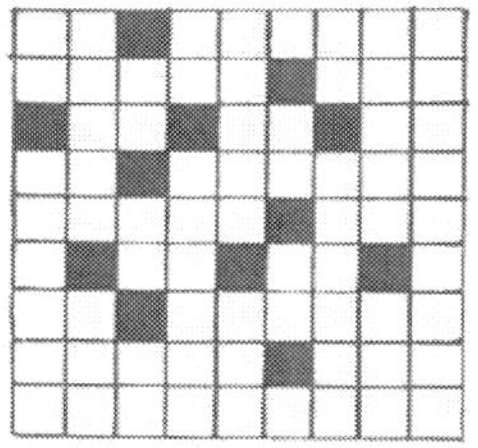

Fig 168

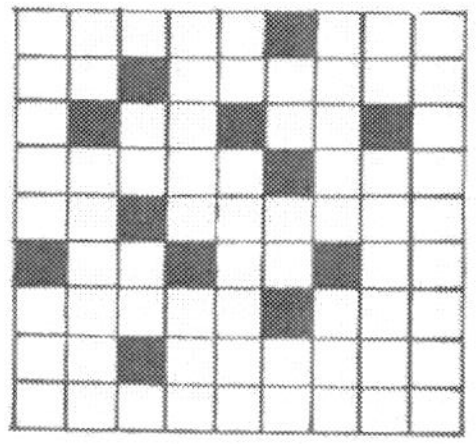

Fig 169

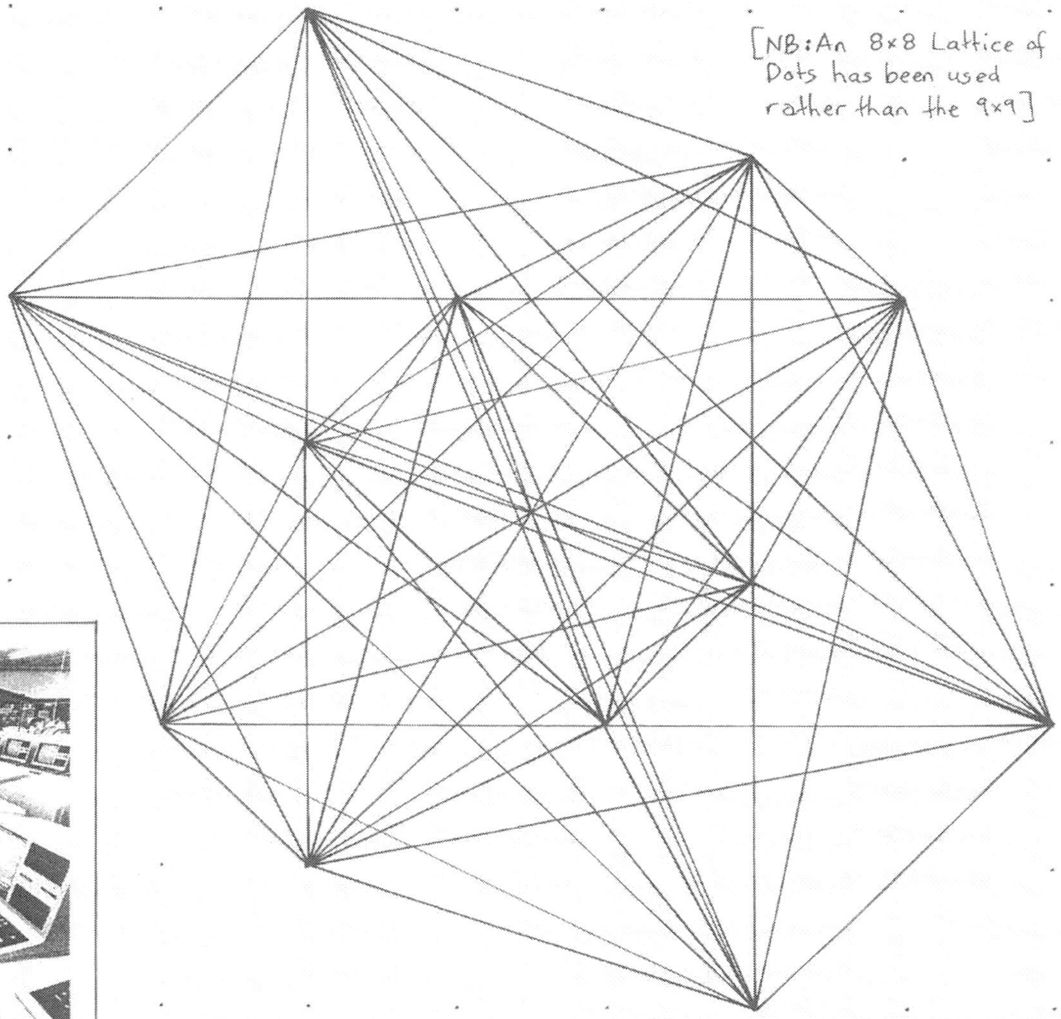

Fig 170

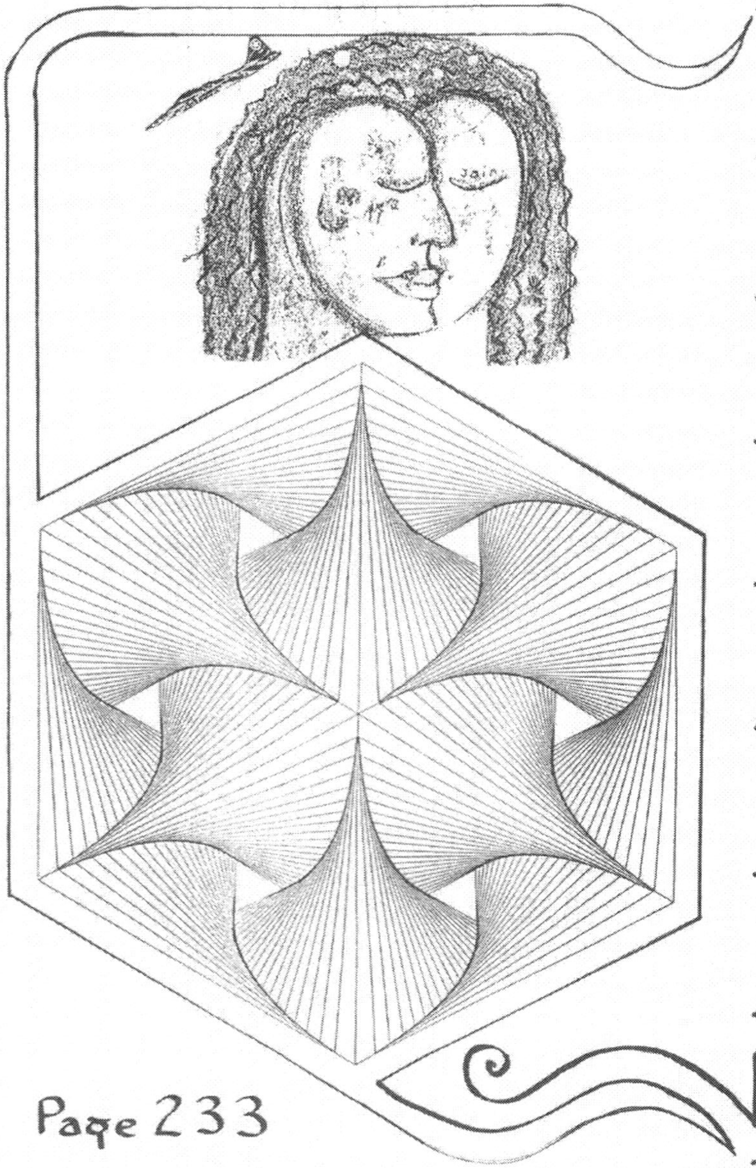

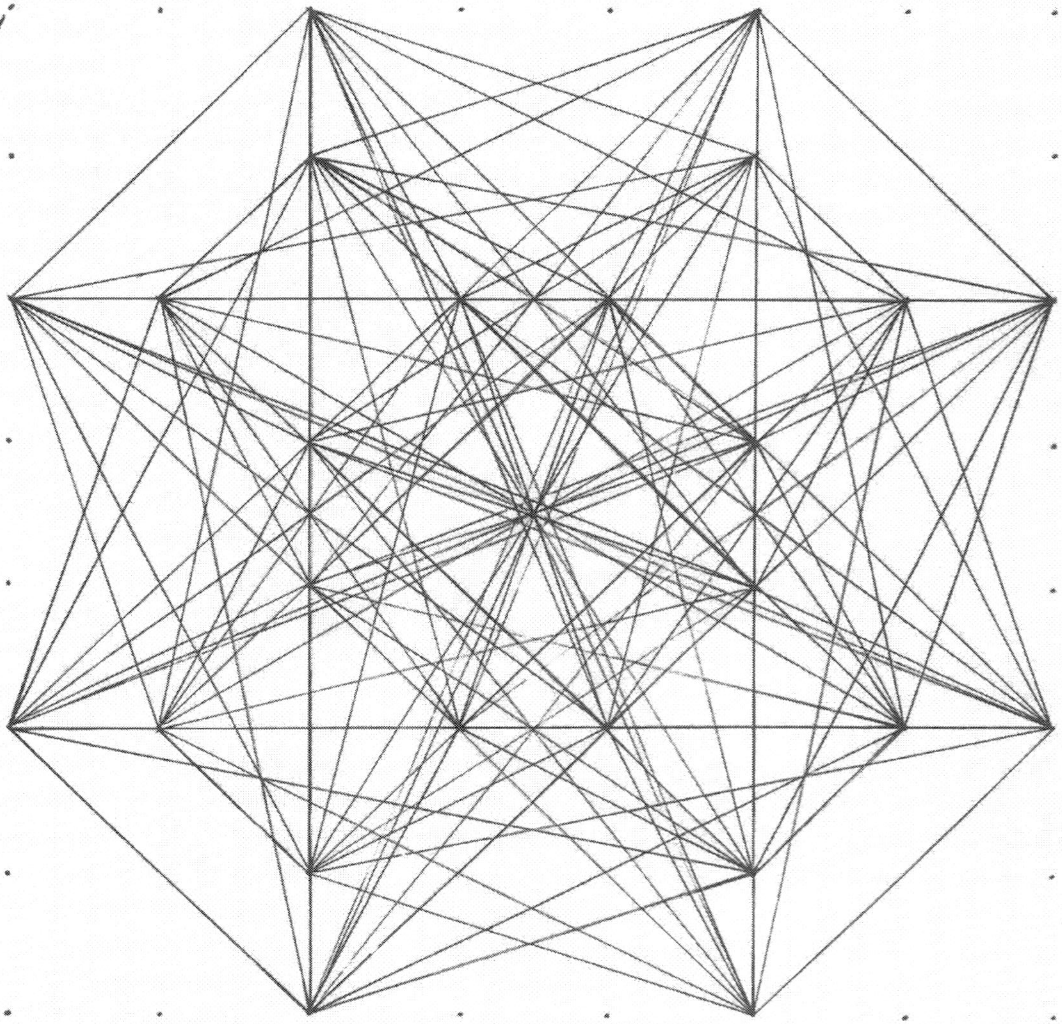

FIG 171

THE BOOK OF MAGIC SQUARES

The regenesis & re-education of the Living Flame Geometries of the magic Square Artforms are a relatively new star in the sky of modern esoteric graphics. As a Rishi, or a Nabi (a prophet, a bearer of a particular message for a particular time), I teach "e pluribus Unum" means One out of Many or Unity out of Diversity, multiplicity... The Greek word 'Kosmos' means literally 'ORDER', thus a study of the cosmos involves a simple study of number. Marianne Williamson in "A Return To Love" says "Spiritual growth is not about becoming more metaphysically complicated, but rather it is about growing simpler". These magic Squares are the reprogramming of life coded in the language of the stars & this is the reprogramming of ascending life energies from the physical worlds of the lower heavens spoken of in the illustrious scrolls of the Ascended Masters who speak of the Father's/Mother's World of many Mansions. M.Sqs. are the many Mansions: The As Above, So Below, As like the Left, so is the Right. A harmonious arranGement of many Paths where all is EQUAL... Inter-Dimensional.

6	1	8
7		3
2	9	4

RISHI:

Meaning researcher, literally, but spiritually, one who has devoted his entire consciousness to the analysis of and experimentation into the nature of man. MAHA (great) RISHI (spiritual researcher) is a common name in India but in ancient times it was reserved only for those who did original work and founded the great philosophical and spiritual systems. MAHARISHI means the same. Einstein was a great Rishi who wrote on the nature of reality as well as physics, although this part of his work is neglected.

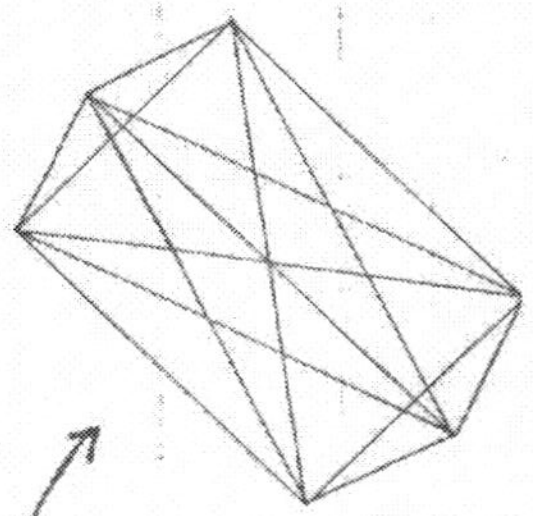

fig 174

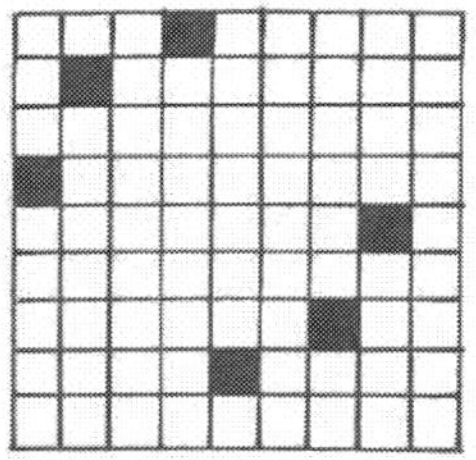

fig 172

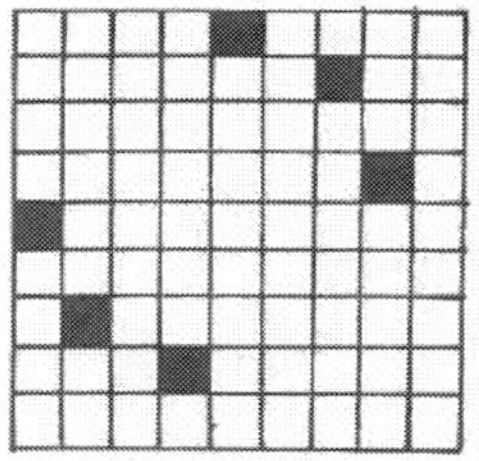

fig 173

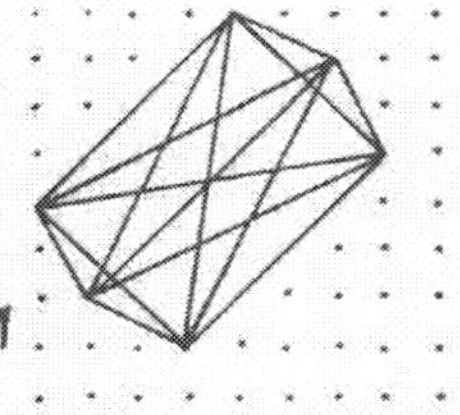

fig 175

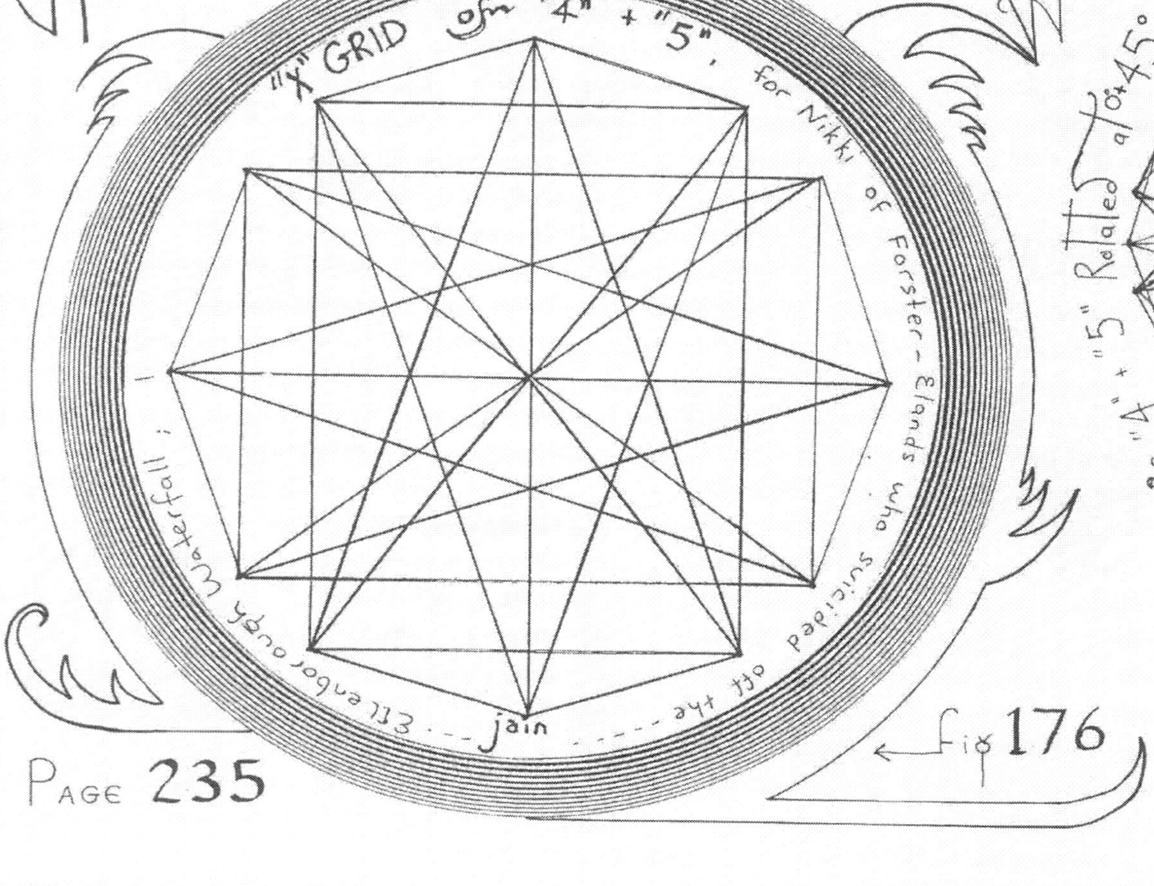

fig 176

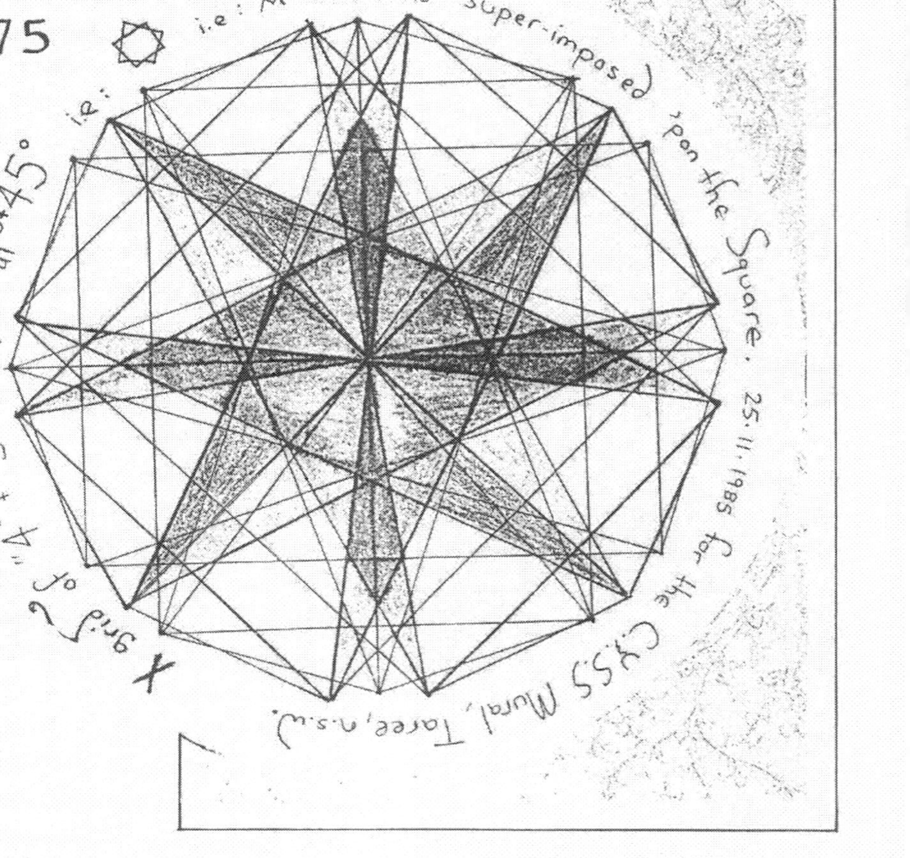

fig 177

FIG 178

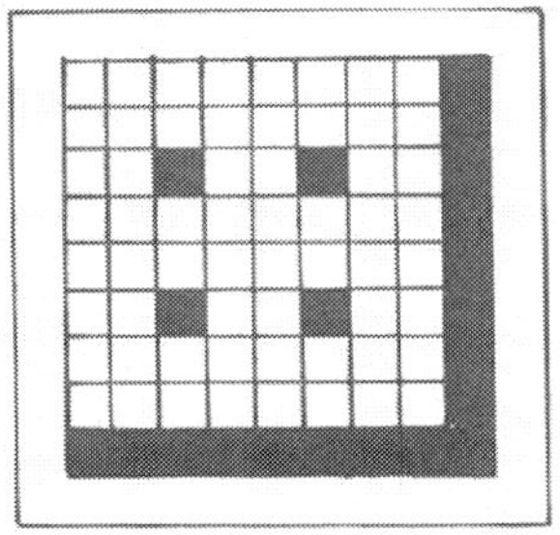

FIG 179

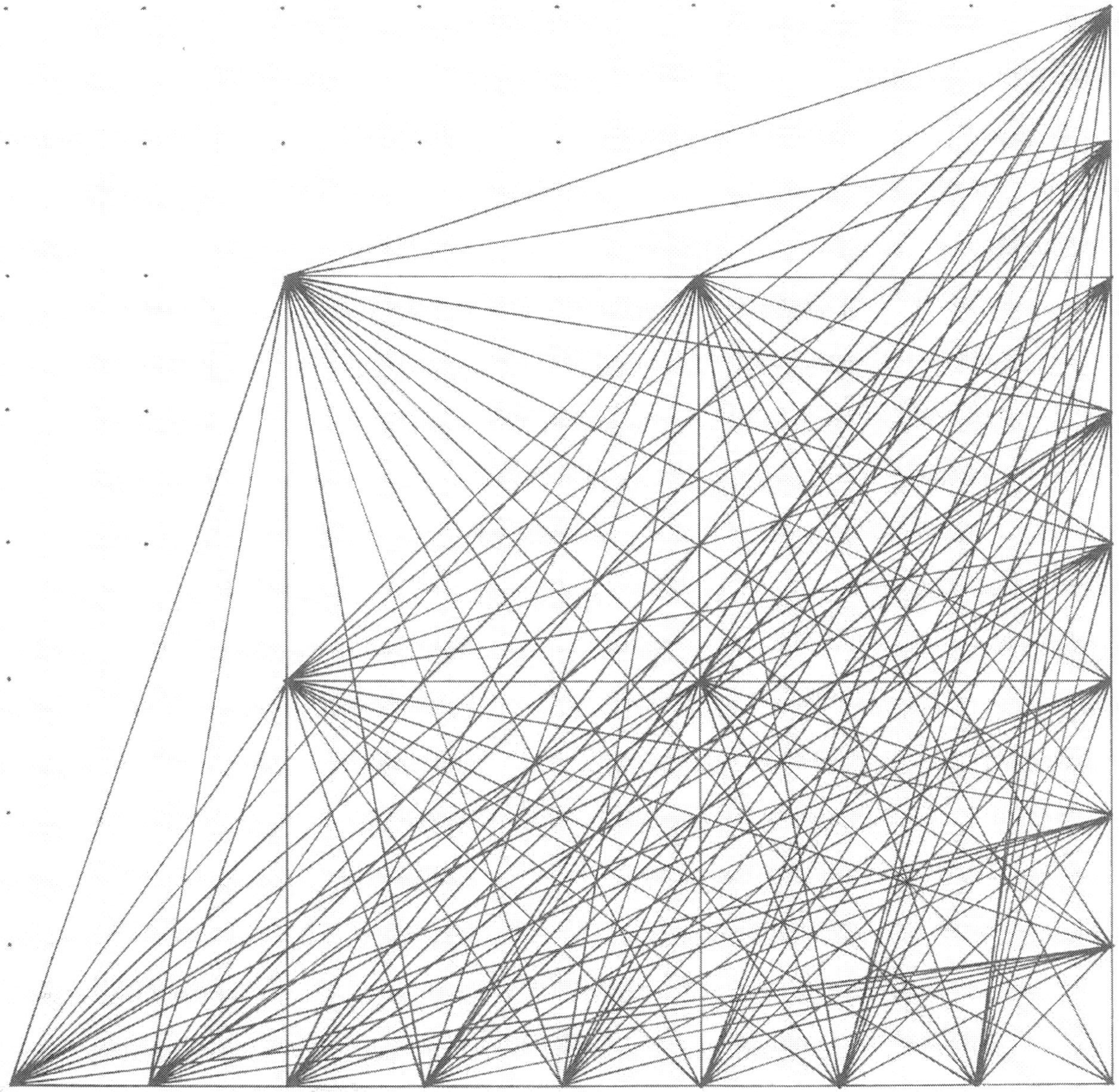

DID YOU KNOW ?

- DID YOU KNOW ? that certain ocean cruisers have painted on their decks huge Magic Squares of 3×3. I first saw this when I was 12 years old on the Russian cruiser "Shota Rustavelli". This was used for scoring in such games known as "Shuffleboard". This was the first magic Sq. I ever saw and it subconsciously changed my Life!
- DID YOU KNOW ? that the MAGIC CUBE of 3×3×3 has all its Columns + Rows + Diagonals in 3-Dimensional space adding up to the Magic Number of 42 ! which is the HitchHiker's Guide to the Answer to the Universe !
- DID YOU KNOW ? that ancient books on Magic attribute Magic Squares to Planetary Energies! By connecting certain numbers of the Hebrew Magic Squares e.g 3×3 at certain hours of the day, according to certain angelic Names, the user could potentize their wishes to come true or manifest.

SATURN

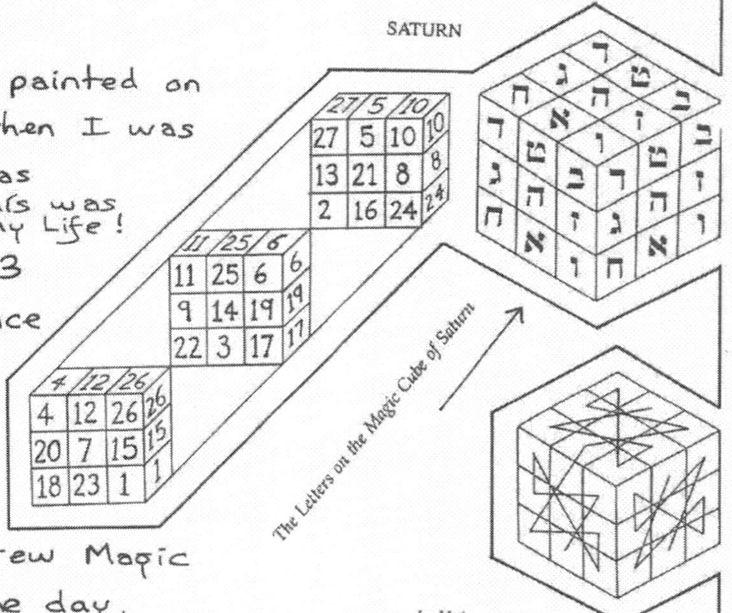

In Hebrew notes

4	9	2
3	5	7
8	1	6

The table of Saturn in his Compass

The Seals or Characters

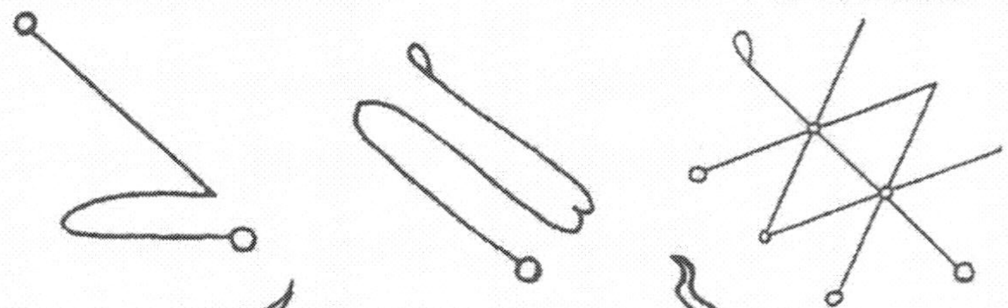

Of the Spirit of Saturn

Of the Intelligence of Saturn

Of Saturn

ד	ט	ב
ג	ה	ז
ח	א	ו

PAGE 237

DID YOU KNOW ?
DID YOU KNOW ? DID YOU KNOW ? DID YOU KNOW ?

◆ If I gave you 9 playing cards from 1 to 9 and asked you to make a M.Sq. of 3×3, your chances of being correct are 1 in 45,360 ! This is the mathematics of it : To make any 3×3 square of numbers (not magic) you have [1 → 9 !] i.e 1×2×3×4×5×6×7×8×9 chances (! = factorial) which = 362,880. But there are only 8 possible solutions shown here ⟶ that exist out of this great number, and therefore the answer is

362,880 ÷ 8
= 45,360

DID YOU KNOW?

6	1	8
7	5	3
2	9	4

2	9	4
7	5	3
6	1	8

6	7	2
1	5	9
8	3	4

2	7	6
9	5	1
4	3	8

These are the 8 possible Variations or Permutations of the Loh-Shu or the Magic Square of 3 x 3. They all have sums of 15.

8	1	6
3	5	7
4	9	2

4	9	2
3	5	7
8	1	6

8	3	4
1	5	9
6	7	2

4	3	8
9	5	1
2	7	6

JOIN - THE - DOTS

THE ART OF MAGIC SQUARES

INTERDIMENSIONAL COMMUNICATION

◆ DID YOU KNOW ? that the first itemised 880 permutations of the M.Sq. of 4×4 were left in tabulated manuscripts by Bernard Frénicle de Bessy, membre de l'Ācadémie Royal des Sciences, circa 1693.

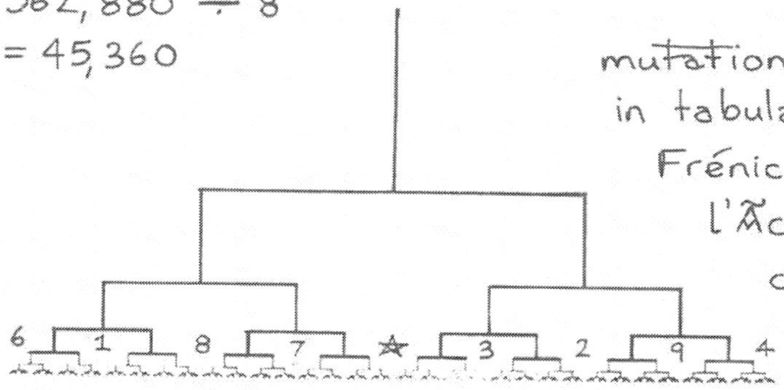

"THE CITY" by JAIN

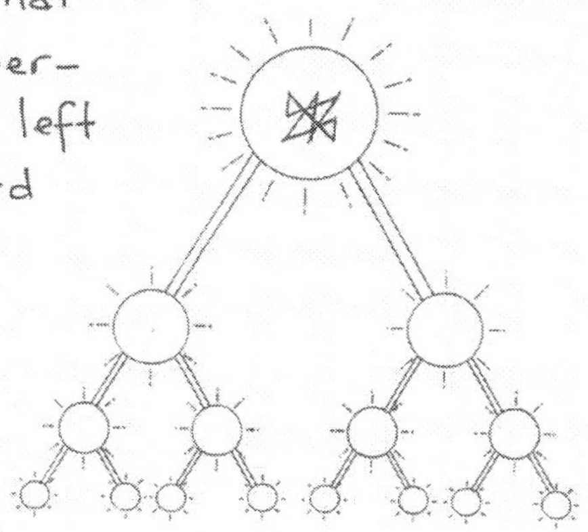

Light-Body Network

BINARY CODES OF 8NESS

DID YOU KNOW ? DID YOU KNOW ? DID YOU KNOW ? DID YOU KNOW ? DID YOU KNOW ?
DID YOU KNOW ? DID YOU KNOW ? DID YOU KNOW ? DID YOU KNOW ?
DID YOU KNOW ? DID YOU KNOW ? DID YOU KNOW ?

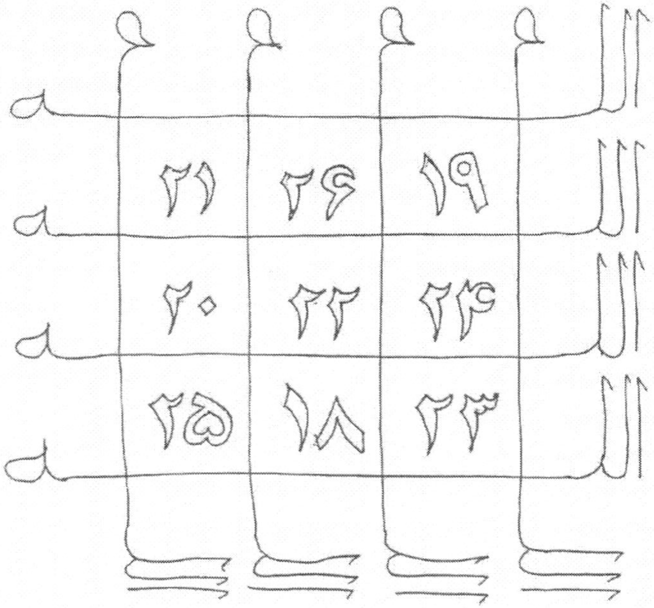

◆ DID YOU KNOW ? that the numerical significance of the Divine Name "ĀLI, The Ālmighty", from which was derived the name of the Son-In-Law of the Prophet, can be determined by adding up the numbers in any complete row and subtracting the 2 numbers left + right of centre.

Name of Āli عَلِىّ = 110

Magic Sum of rows etc = 330.

109	114	107
108	عَلِىّ	112
113	106	111

◆ DID YOU KNOW ? that the Divine Name ĀLLĀH corresponds to 66. The Square itself is formed of the Name in creative calligraphy, containing thrice three numbers which in any direction add up to 66.

21	26	19
20	22	24
25	18	23

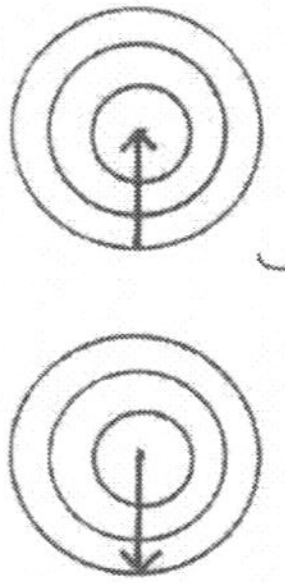

IMMANENCE : indwelling. God lives in the Church.

TRANSCENDENCE

The Goal of SUFISM is to gather all multiplicity into UNITY. Mohammed said of himself : "I Am AHMAD without the "m" (Ahad means Unity). I Am an ARAB without the "a" (Rabb means Lord). (Mohammed The Prophet channelled The Koran).

DID YOU KNOW ? DID YOU KNOW ?

that according to two ClairVoyant Scientists : W. Leadbeater + Annie Besant , from about 100 years ago , remarkable + accurate drawings of all the Elements, including newly discovered Isotopes, were observed + published via the Theosophical Society. Shown here is the atomic structure of Diamond Lattice ⟶

Diamond lattice, according to micro-psi observation

The Magic Square of 3 x 3 Tiled 81 times at 0° and 90° Alternations

which needs to be compared to the tessellated or tiled Magic Square of 3 x 3 that is shown magnified + magnificently here plus also on p 57 Vol 1 and was constructed on p 55 Vol 1 as Fig 35. "MICRO-PSI" means that as Siddha-Yogis they had the Divine Power to miniaturize themSelves + enter the world of atoms/quarks + galaxies ! One of the pair entered this observable dimension + the other rendered the Atomic Art-work. Notice the incredible similarity of the two drawings. Science meets Mysticism/Art. Magic Square ArtForms , I therefore ∴ conclude, are like a secret backdoor entrance into the WORLD OF ATOMIC PHYSICS. Let it enter the school maths Curriculum!

DID YOU KNOW ?   DID YOU KNOW ?   DID YOU KNOW ?

that the SECRET to understanding

SACRED GEOMETRY

is to visualize 2-Dimensional objects in their true 3-D or HOLOGRAPHIC state i.e A circle = a sphere, A triangle = a tetrahedron, a square = a cube, a ✡ = a star tetrahedron, a ★ = a dodecahedron. In this regard, we can now see 2-D magic Squares, not only as Magic Cubes but as toruses (Fig A), which is like an imploding sphere; as HyperCubes (Fig B) which is a 4th Dimensional cube or a Cube within a Cube; as Octahedrons (Fig C) which is the 3-D form of a diamond shape ◇; as Cubes (Fig D) which shows opposing pairs in balance i.e 2+8 = 10, 1+9 = 10 and 3+7 = 10. Both Figs C + D are the M. Sq. of 3x3 or Lo-Shu seen in an exciting view.

Fig A is by J. B. Rosser + R. J. Walker of Carnell University in 1938 which joins top + bottom row to form a cylinder which stretches + bends into a torus! All closed loops now = 34 in any direction. If we start at any cell and move two squares away in any direction along a diagonal we always arrive at the same cell. This cell is called the ANTIPODE of the cell where we began. Every pair of antipodes on this diabolic doughnut does total 17. Similarly, Fig B, the HyperCube has pairs summing to 17 which all pass thru the Centre. A Diabolic Square remains symmetrical even when a column or row is transferred from one Position to another!

Fig D

Fig C

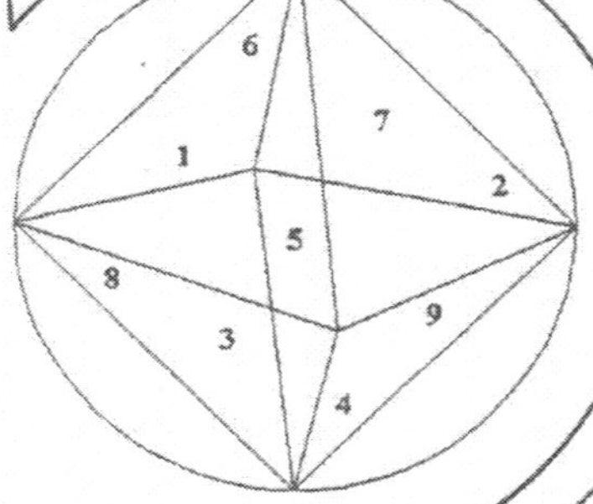

7	12	1	14
2	13	8	11
16	3	10	5
9	6	15	4

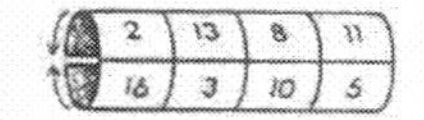

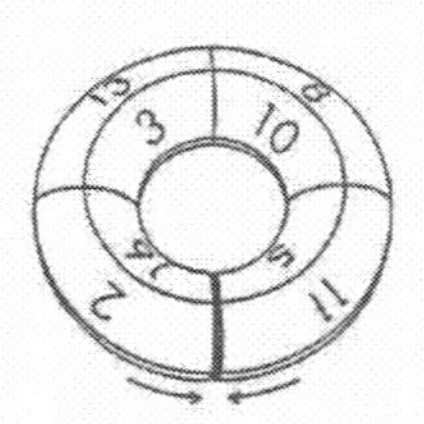

The diabolic doughnut.

Fig A

1	8	13	12
14	11	2	7
4	5	16	9
15	10	3	6

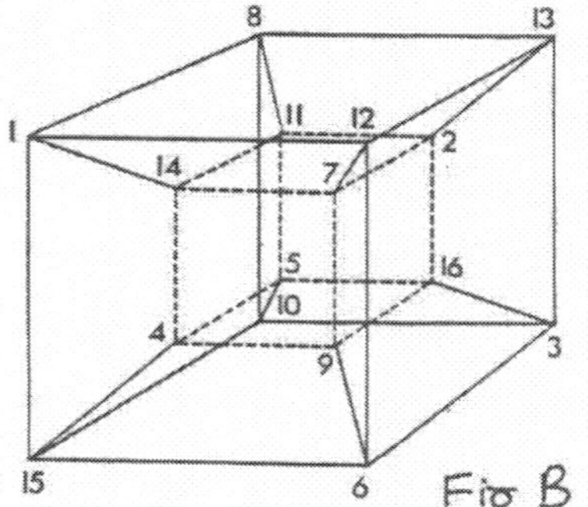

Fig B

Diabolic hypercube and one of its 384 diabolic squares.

DID YOU KNOW? that CLAUDE FAYETTE BRAGDON, a prominent U.S. architech + occultist who died in 1946, was fascinated by his discovery that on most magic squares a line traced from cell to cell in serial or sequential order will produce an artistically pleasing pattern. Other patterns can be found by tracing only the odd or even numbers or cells. Bragdon used "magic lines" obtained in this manner as a basis for textile patterns, book covers, architechural ornaments, and the decorative chapter headings of his autobiography "MORE LIVES THAN ONE". His design for the ventilating grill in the ceiling of the Chamber of Commerce in Rochester, New York, where he lived, is derived from the magic line of the Lo-Shu.

DID YOU KNOW? that the EGYPTIANS used the magic square of 3×3 → or Lo-Shu for Crystal-Gazing. The Square was drawn in white chalk on a black cloth with a crystal placed in the central cell of 5. For the subconscious to reign free or travel in absolute time, the crystal-gazer needs to dislocate the busy mind or suspend all conscious mental activity by tuning into the crystal ball.

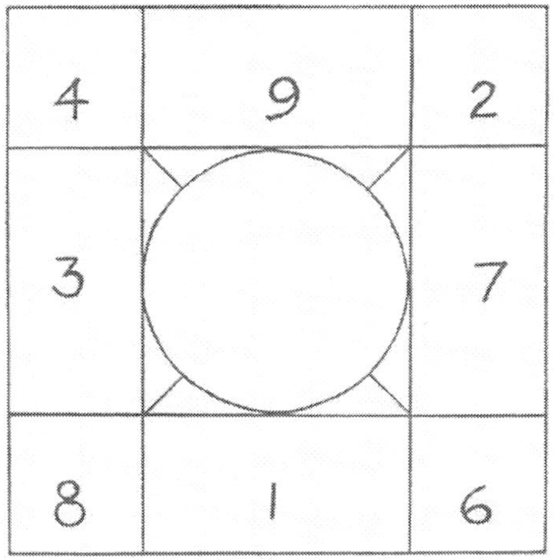

DID YOU KNOW? that MULTIPLICATION SQUARES are formed when the cells of a magic square are not added but multiplied. Shown here is a most ingenious magic square of 3×3 using the numbers of the Lo-Shu as power numbers or indices eg: 2 to the 6th power = 2×2×2×2×2×2 = 256 etc and multiplication of any 3 numbers eg 256 × 2 × 64 = 32,768 which is the magic sum of multiplication for the 3 rows, the 3 columns and the 2 diagonals.

2^8	2^1	2^6
2^3	2^5	2^7
2^4	2^9	2^2

=

256	2	64
8	32	128
16	512	4

DID YOU KNOW? that the MASONS used magic squares to convey cryptic messages. The Mason's Mark shown here linked 8 of the 9 cells of the Lo-Shu to indicate that the Eight (of Balance) is the way to the Nine. In Arabic, Nine represents the letter "TA" whose hidden meaning is "secret knowledge". On another level, the shape of this mark which looks like the number four with a bar (or an attached curve) roughly resembles the Arabic word "HOO" the sacred dervish liturgical word, which when chanted produces ecstatic states.

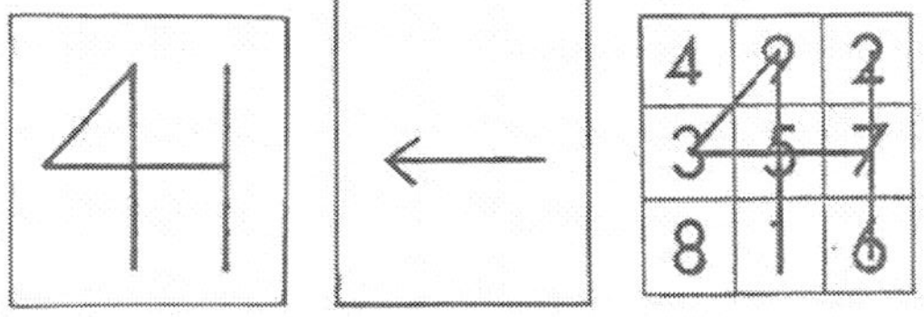

DID YOU KNOW? that the 3×3 M.Sq. was engraved on silver to ward off effects of the bubonic plague.

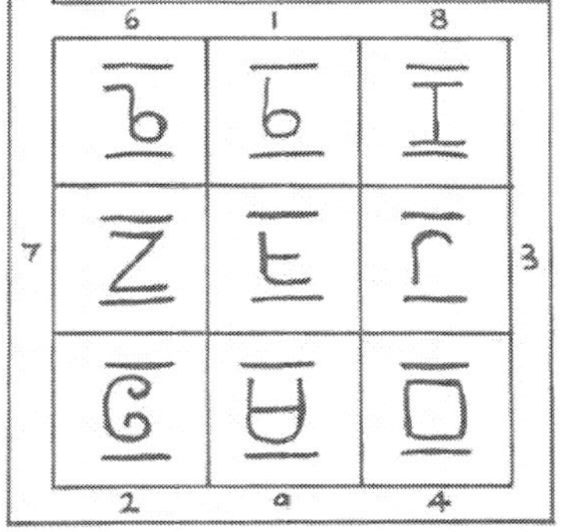

← An ETHIOPIAN M.SQ of 3.

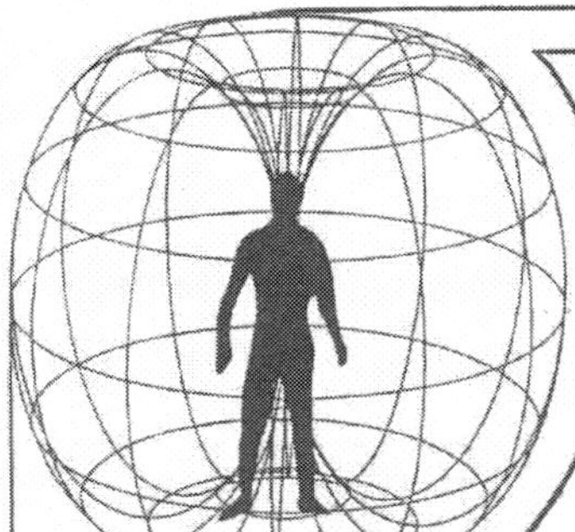

Seven-point Star

Eight-point Star

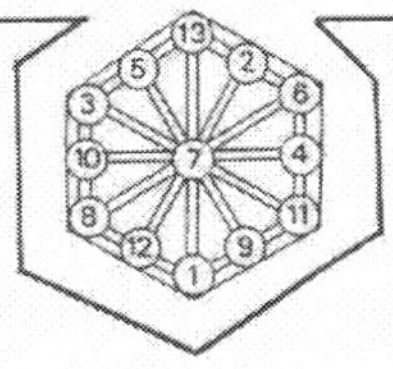

Magic Hexagon

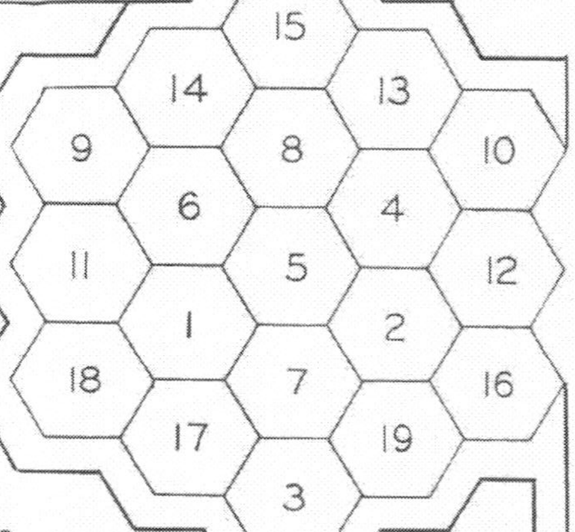

DID YOU KNOW ? DID YOU KNOW ? DID YOU KNOW ?

● that Magic Squares are ASTRO-PHYSICAL TEMPLATES & will be used by future technologies for Free Energy Systems. Professor. J. SEARL of England has already demonstrated how the primary pairing harmonics of M. Sqs, (ie the Balance of the Pairs) related to alchemical sequences, create vertical lift-off for U.F.O. crafts. An understanding of M. Sqs + the Periodic Table of the 96 Elements is important. (U.F.O.s are mainly man-made Inner Earth crafts!)

● magic Squares are molecular crystal structures used for the induction of cosmic rays and for multi-Dimensional LIGHT COMMUNICATION.

● Did You Know that the basic formula for the 2-Dimensional Magic Squares' Magic Sum or Constant $C = n/2\,(1 + n^2)$ is replicated in The Higher Dimensions. For example: in 3-Dimensions, the magnitude of the Magic Square (n) becomes the Magic Cube where $C = n/2\,(1+n^3)$. By substituting n=3 into this formula we get $C = 3/2\,(1+3^3) = 42$. Then the next step is crucial as it takes us into the 4th Dimension. Can you see a pattern. In 4th Dimension $C = n/2\,(1+n^4)$. By substituting $n = 3$ into this formula we get $C = 3/2\,(1+3^4) = 123$. Thus, if someone wanted to know what the magic frequency is for the 5th Dimension using the simplest M. Sq. of 3, then you would substitute $n=3$ into $C = n/2\,(1+n^5) = 3/2\,(1+3^5) = 366$. Happy exploring!

● Did You Know that these Magic Square CONSTANTS are the true indexing of the vortical spin rate of the chakras + higher dimensions. These spin-rates equate to our ASCENSION process or the transformation of the human bio-crystalline system by guiding the circuit of pure energy emissions within the geometry of geometries. Thus you may view these Magic Squares, Cubes, Stars, Hexagons, Crosses + Magic Triangles as a vibrational mirroring or a visual analog of electron patterns that open dimensional gateways. The centre of a Magic Cube is a Vortex or a Torus Doughnut that carries the Divine Programming for your INVOLUTION to remember the essential components for riding your MERKABAH Vehicle back Home to the

STARS.

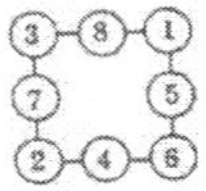

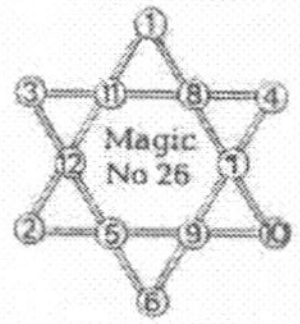

Six-point Star

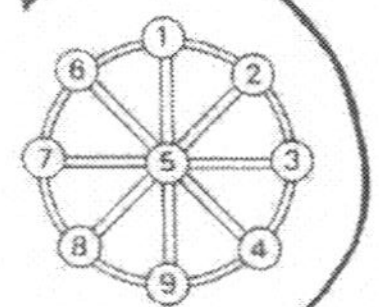

Magic Wheel

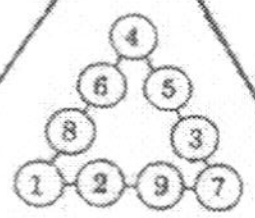

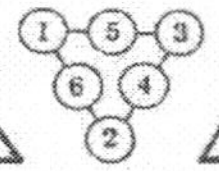

Magic Cross

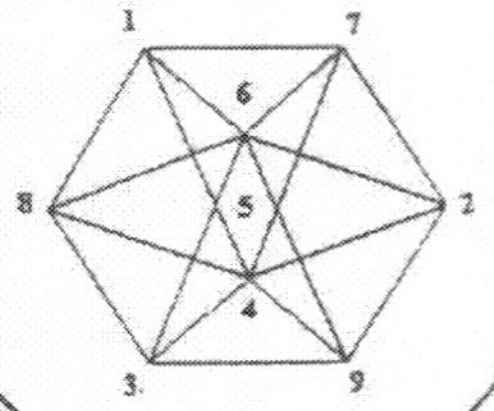

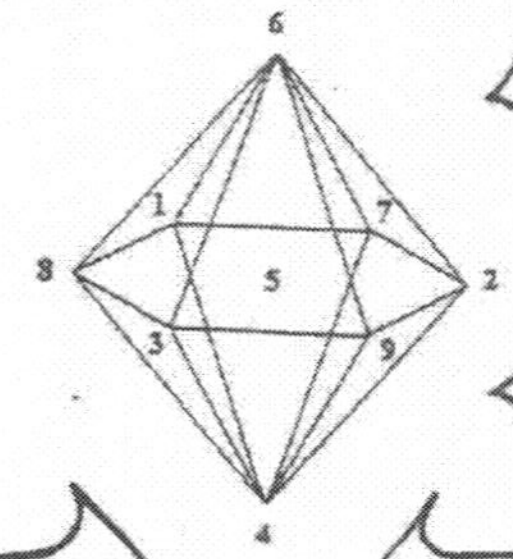

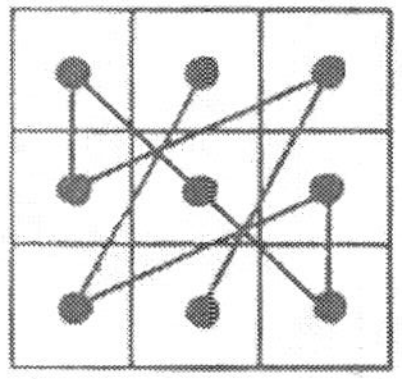

Ref 1

References To Majik Square Graphics From Other Sources

DID YOU KNOW ? DID YOU KNOW ?

Reference 1 is from "Islamic Patterns" by Keith Critchlow p 44 linking the numbers of the Majik Square of Three (M.Sq of 3).

Reference 2 was found on the backcover of one of Claude Bragdon's books illustrating another one of his books known as "Projective Ornament". It appears to be a M.Sq of 7.

Ref 2

Reference 3 is from a German book "Mathematische MuBestunden" by Prof. Dr Hermann Schubert p217 Berlin 1941. It appears to be a M.Sq of 8 in that all its columns + rows add up to 260 but it fails in its two diagonals. Though it is not a true majik square the diagram shows the principle of transforming number patterns into an ArtForm by drawing a long continuous line from alpha (α) the beginning to omega (Ω) the end.

It has been my vocation to go to librarys thru-out the world, especially in Australia, and fotocopy any reference of any literature on this specialized topic of majik squares also known as Harmony Squares. On this page I have listed the only three recent and isolated findings in 1986 at the Sydney University. I put my name to this Encyclopaedia of Majik Square Graphics or Yantras to acknowledge that it is an inspired and independent production. It has not been copied from any other book. There is no other book or source on M.Sq Graphics to my knowledge, but intuitively I know, in some dusty library, there is. This page is to give credit to the three above-mentionned references. May this book plant the Seeds for the Lost Knowledge of our ForeParent's Cosmik Designs to sprout forth.

Ref 3

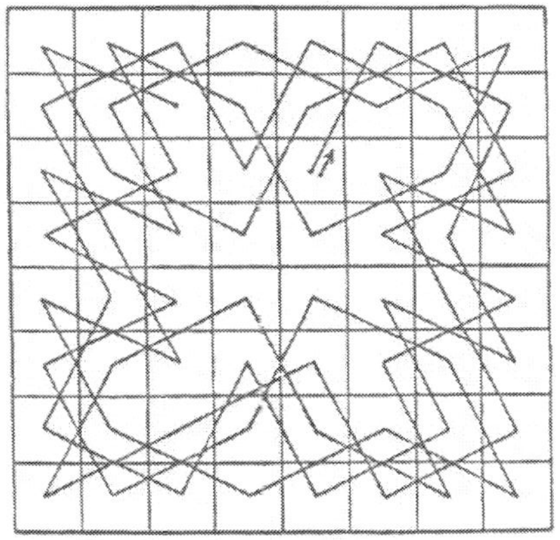

« THE BOOK OF HARMONY SQUARES »

THE BOOK OF MAGIC SQUARES

6	1	8
7	5	3
2	9	4

VOLUME ONE.

DID YOU KNOW? ... DID YOU KNOW that the centre of the Sino-Tibetan Zodiac or Calendar of Animals is The Magic Square of 3 x 3. The 8 numbers around the central cell 5 ... represent the Eight trigrams shown here as full (Yang) or broken lines (Yin). This is also shown on p26 + p60. This diagram is a clear linking of Magic Squares and TIME codings.

All of the ancient principles of FENG SHUI: the Art of Aesthetic Placement are based upon these Numbers.

• DID YOU KNOW ? that the Lo-Shu or ancient M. Sq. of 3×3 can be formed by swapping around certain numbers of the simple Natural Square of 3×3.

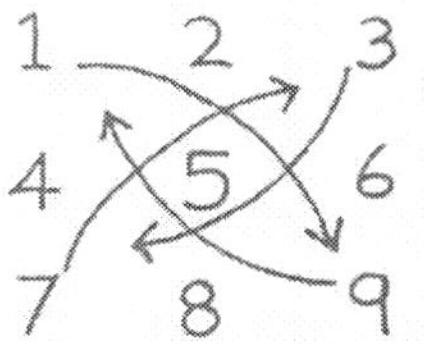

• DID YOU KNOW ? that in the Far East (India) the M. Sq. of 3 is worn as an amulet (JUNTRA), made of copper, upon the biceps, for spiritual strength.

• In certain villages in India, like Pushkar of Rajasthan, all buisnesses + shopkeepers have in their alcoved altar or small wall recesses, framed diagrams of M. Sqs. written in Sanskrit, to invoke success for the day. This is known as PUJA : lighting of incense, candles + offerings of fruit + mantras (potentised sonic prayers).

• DID YOU KNOW ? that in 1838 a French work on the subject of M. Sqs. ran to three volumes of compilations.

• DID YOU KNOW ? that the earliest recorded 4×4 M. Sq. was found in an C11th or C12th inscriptions at Khajuraho, India.

• DID YOU KNOW ? that the central cell of the Magic Cube of Eleven (involving 11 × 11 × 11 numbers) can only be 666 !

• That M. Square patterns can be created in different ways to form sacred symbols. eg 1 : The Star of Solomon ✡ is formed, not by joining 1 → 2 → 3 etc in sequential order, but by a Row-by-Row analysis as explained in "The Book of M. Sqs" Vol. 2 pp 149 + 226. Essentially, in 3-Dimensions this is The Merkabah + Silicon atom ! eg 2: M. Sqs. can be converted to sine-wave like patterns/signatures, again by using a Row-by-Row analysis as explained in Vol 2 pp 24, 25 + 200.

MAGIC NUMBERS

The 3 × 3 Magic Square in which the numbers 1 through 9 were distributed is the earliest known of this type of problem, and has been traced back as far as fourth-century BC China. To the ancient Chinese, this *Lo-shu* represented either the universe or China, and its middle number therefore represented either the center of the universe or the Emperor – in Chinese eyes these were effectively the same thing. Chinese mathematicians went on to construct 5 × 5 and 7 × 7 Magic Squares of various types, although they found it a little more difficult with 4 × 4 and 6 × 6 Squares and eventually came to regard these as ominous.

It was for its 'lucky' properties that the 3 × 3 Magic Square became popular in the Arab world in the tenth century AD. The associations of such Squares with the universe, introduced contemporaneously, quickly lent them religious interpretations, as expounded in an encyclopedia published in about 989 by the Brotherhood of Purity. Sufi mystics saw the Squares as representations of Life in constant motion, renewed and rejuvenated by the source of power at the heart of the universal design.

The Hindus of medieval India regarded 4 × 4 and 6 × 6 Magic Squares as particularly potent and lucky, and went on to construct larger Squares of even-numbered sides, eventually establishing a sophisticated methodology for doing so.

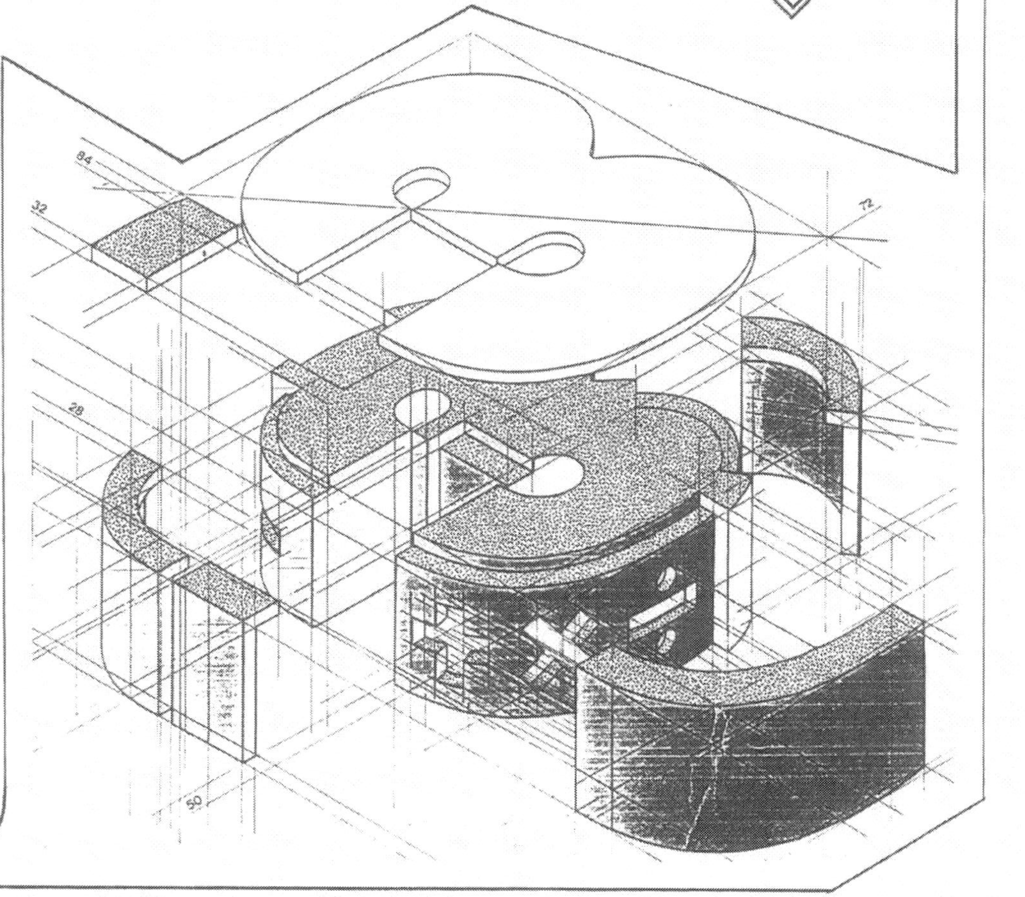

ACKNOWLEDGEMENTS of GRAPHICS USED FOR VOUME 3

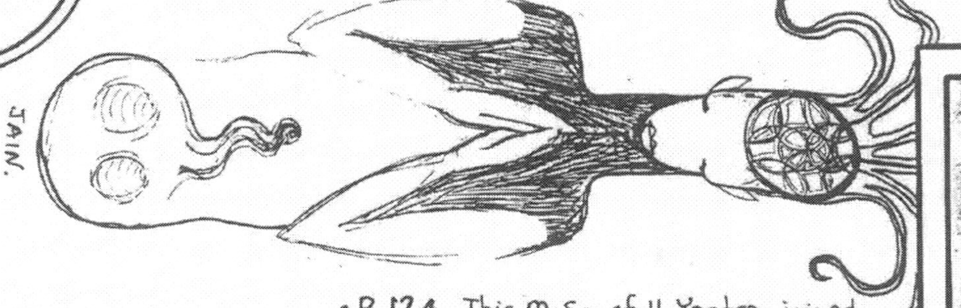

ACKNOWLEDGEMENTS : CHAPTER 11

• P 124 This M.Sq. of 11 Yantra, joined side by side, was first published in "The Time Twins" by Jain 1981 (a.k.a Enfant -de Neptunuu.
• P 126 a) All the text is from the Editors Brief Commentary from "The Gospel of the Holy Twelve" (This is a rare book in my M.Sq. Library as I have not seen another like it in 15 years. b) LHS Drawing by Jain.
• P 127 a) LHS Drawing by Jain. b) Top. RHS & Bot. RHS by Kahlil Gibran whose quaint home I visited in the Lebanon Snowy Mountains. Being myself of Lebanese descent, I quite relate to this artist/author who said things like "Your Children are not your Children..." etc.
• P 128 Top Cent. Drawing by Jain. • P 129 LHS Drawing by Jain.
• P 130 Top RHS M.Sq. 11 Yantram fotocopied from a photo of a 2m x 2m multi-colored cotton mural by Evelyn Krosby 1985 Kippaxs Studio N.S.W.
• P 131 Bot LHS Horse from an instructional Sumi-e book that I once studied from, title unknown. Art with meditation & with one pure stroke.

• On this page a) Bot LHS border : book binding, Renaissance. b) text within from "Down To Earth" mag Brisbane 1995, author unknown. c) RHS Drawing by Jain.
• P 132 Top LHS typed text & Bot RHS "11 : 11" from the teachings of Solara who visited Byron Bay. • P134 This Mohammedan dome is an intricate array of inter-connecting 11-rayed stars & 9 rayed stars. (Country of origin unknown). • P 135 a) Top LHS Drawing by Jain. This illustration of Jesus was painted wildly with treacle, cigarette ash, & coffee stains on Thursday Is. 1982 (I lived in rainforests for 2yrs with no money & just picked fruit off trees). b) Bot LHS Quote, Azena Ramanda, chanelling St. Germain.

Medifocus is a special humanitarian monthly membership activity with which each Rosicrucian is acquainted. On the first Sunday of each month, at any hour you select, you will enter into a five-minute period of meditation, focusing your thought upon a specific troubled area of the world. The part of the world you select will depend on which troubled area is particularly significant to you as an individual. This may change from month to month, or it may remain important to you for a longer period of time.

The Rosicrucian Order is *not* a political organization. The basic purpose of *Medifocus* is a humanitarian effort directed toward *world peace*.

ACKNOWLEDGEMENTS : CHAPTER 12

NXXDXD VXRY MUCH!

Xvxn though my typxwritxr is an old modxl, it works quitx wxll xxcxpt for onx kxy. Thxrx arx 46 kxys that function wxll xnough, but just onx not working makxs thx diffxrxncx.

Somxtimxs it sxxms that our group is somxwhat likx my typxwritxr, not all thx kxys arx functioning propxrly. You may say: "Wxll, I am only onx pxrson. It won't makx much diffxrxncx." But, you sxx, thx group to bx xffxctivx nxxds thx activx participation of xvxry pxrson.

So thx nxxt timx you think you arx only onx pxrson and that your xffort is not nxxdxd, rxmxmbxr my old typxwritxr and say to yoursxlf, "I am a kxy pxrson and am nxxdxd vxry much."

• P 136 Border of 22 circles by Jain. Each circle is 12-sided with increasing-sized squares forming 1 : ·618 helices.
• P137 a) RHS 12 overlaid circles, computer generated by Nic Roche of Byron Bay 1991, his logo, silk-screened. b) LHS "The Circle" by Jill Jackson, Aust'n Photography mag April '79.

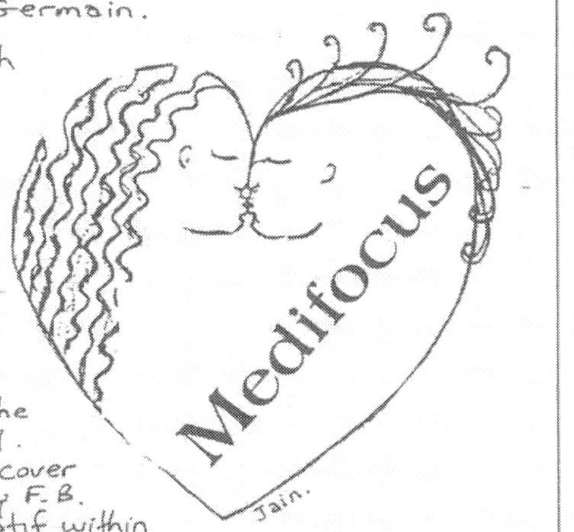

• P138 a) Cent. Star is from "The Secret Symbols of the Rosicrucians" from my friend John Opit's magick library in the rainforest/mountains NSW 1991. b) Top LHS star from the front cover of "The Enigma of Israel" by F.B. Folbrook. The small circular motif within its centre is a Mosaique, art gréco-romain Ostie, Italie. c) Bot. Cent. Arab motif. d) Top RHS drawing by Jain '82. e) Cent. Spiral Star, New Logo for Raël intergalactic movement. f) The Seal of Solomon ✡ refers to the teaching of Hermes by showing the world above & the world below. Hermes = Thoth the builder of the Egyptian Pyramids + creator of our languages
• P 139 LHS Drawing by Begaye 1981
• P 140 a) Bot LHS Drawing by Jain b) Bot. Cent. The Swaztika in Star of Solomon was the original logo for Raël, spokesperson for E.T. Embassy on Earth, but because of its controversial nature (Jewish + Nazi symbols) it was kindly replaced by the Spiral Star on p 138 c) RHS text from "Islamic Science" by Seyyed Hossein Nasr 1976. World of Islam Festival Pub. Co. Ltd.
• P 142 Bot LHS Drawing by Platpus ('Radiance') of Wytaliba.
• P 146 This 12x12 Knight's Tour first appeared to me in a German book of m. Sqs. "Mathematische Mußestunden", Prof. Hermann Schubert.

Man's Higher Vehicles
A more exalted view of man relates him also to the world above the zodiac. With his 'supernatural' faculties of Reason *(Ratio)*, Intellect and Mind *(Mens)*, he soars through the triple world of the angelic hierarchies and contacts God himself. These higher faculties are, of course, outside the physical and etheric bodies, whose domain stops at the circle which the man touches. Beyond that, he is on the way to becoming 'Universal Man' and eventually a God in his own right.
UCH II, a, 1, t.p.

• On this **Page** illustration from the Robert Fludd material 1619 →

• **P 147** The Vision Mandala poster from 1980s alternative lifestyle festivals. The "All One Family" gatherings are seasonally & open invitations for those who do not smoke, take drugs, alcohol nor eat meat. Cent. inscription reads: "Healing the Earth, Healing Ourselves, Healing each other". Initiated by my friend Dik Freestun.

• **P 148 a)** LHS Triangle in Circle's Centre + Bot LHS star with symbols are from "The Gospel of the Holy Twelve" a rare Essene bible in my library. **b)** Cent. Drawing by Jain.

• **P 149 a)** Top RHS from "The Book of Charms & Talismans" by Sepharial. **b)** Top LHS image from "Crystal Visionary" book Vol 1 1988 Michael Ney. **c)** Bot LHS from an illuminated copy of the Koran, influence of Byzantine art (the construction lines) on that of Arabia (the filling).

• **P 151 a)** Top Cent. A cathedral rose window design in Melun, near Paris, a grand 6-sided masterpiece of light & geometry. **b)** Bot RHS Drawing by Jain. **c)** Bot LHS star has at its centre the all important "Flower of Life" symbol which secretly contains the Tree of Life & the nesting of the Platonic Solids. **d)** Bot RHS from "Spiral Symmetry" by Pickover. Spiral by Keith Wicks. Also see P171 Bot. LHS.

• **P153** Fig 133b is composed of 19 Circles. It is important, for a true understanding of MerKaBah for you to construct this yourself. Fig 133c Michelangelo's famous drawing shows that he & his teacher Vitruvius had an understanding of these 4th Dimensional tools for inter-dimensional travel. Learn to visualize the Star of Solomon 3-Dimensionally as an intelligent auric field around you. This shape is called a Double Tetrahedron or Stella Octangula or more commonly known as the STAR TETRAHEDRON.

• **P 154 a)** Top RHS + Top Cent. are the symbols for Medifocus, from the Rosicrucian Digest Aug 1985 **b)** Bot RHS Circle of 12 12-pointed stars represent the position of the Earth in the 4 faces of God which can never turn around. **c)** Top LHS Fig 135 & the 2 Bot. LHS diagrams are from "Beyond Pyramid Power" by G.P. Flanagan 1975.

ACKNOWLEDGEMENTS : CHAPTER 16

• **P158** tiled border design of 16 15-rayed daisy forms, by Jain.

• **P159 a)** Figs 137 + 138 first appeared to me in W.S. Andrew's "Magic Square & Cubes" 1917. **b)** Top LHS Oval with 16 rays is the Raja Yoga symbol. They wear white robes, practise celibacy & open-eyed meditation.

• **P160 a)** the 16-pointed symbol on RHS of moon is an ornamental motif from Eridou, Mesopotamie, 4me millénaire av J.-C. It also spells the sacred word OM, from "Ornamental Design" by Claude Humbert. **b)** drawing by Rahasya 1978 (Amor y Vida Patty).

• **P 161 a)** LHS Drawing by Jain. 1982 Torres Straits, where I lived for 2 yrs. **b)** Top LHS 16-rayed star is the symbol for Macedonia, now a province of Yugoslavia. **c)** Bot LHS Hand with astrological correspondences set on a B. Franklin M.Sq. of 16x16 from "Encyclopedia of the Unexplained".

• On this Page **a)** Top RHS Sunflower (disc phyllotaxis with sigmoidal growth), from "Spiral Symmetry" by Pickover. Really, it simply has 34 anticlockwise spirals and 21 clockwise spirals. These two numbers of course belong to the ancient Fibonacci Sequence: 0, 1, 1, 2, 3, 5, 8, 13, 21, 34, 55, 89, 144 which is the ultimate mathematics of Nature & Atomic structure.
b) Bot RHS Drawing of a naked woman with a M.Sq. of 3 bum tatoo, by Jain.
• **P 162** LHS Drawing by Jain in Nth India, depicts Reenah Sun, a.k.a. Sambodhi holding our baby daughter on a rock after swimming in McLeod Ganj waterfall pool, 1989. I met the Dalai Lama's personal Astrologer in Dharamsala, nearby, (where most of the Tibetans have flocked to since the Chinese invasion (genocide) & he psychically gave me permission to disseminate the M.Sq. codes / harmonics to the West. NB: The Centre of The Tibetan Calendar / Cosmology is the M.Sq. of 3 & is the Source of this Lost Knowledge to be revived. This year 1995 I "re-discovered" that its yantra produces the atomic structure of beryllium, Star Tetrahedron in 3D & is the basis of the MerKaBah, our Spiritual Light Vehicle!
• **P 163 a)** LHS Drawing by Jain **b)** the wheel nearby is a symbol for Spiritual Energy used by the Reiki Healing Group. • **P 164 a)** Text for RHS from "A Pilgrim's Guide to Planet Earth" 1974 "Spiritual Community" publishers.
b) LHS Jimi Hendrix, whose sonic genius changed my whole universe.
• **P 165 a)** Top LHS seated is a 4 yrs young boy Jerome Gerard who lived with me in Uralba Nth N.S.W. at the time of writing "The Book of Harmony Squares" Vol One 1988.
b) far Bot. RHS of the bedsheet banner is a personal Indian icon that I often had near all my paintings, like a spiritual Overseer of my Art. His name is Danvantari, holding a bowl full of the Nectar of Immortality, the incarnation who brought medicine, from Hari Krishna material. This M.Sq. Tapestry 2m x 3m is Mingkah Sun's Birth Banner. It was hung at Bundjalung beach seashore area where she was born in a tent by the beach, originally meant to be ocean borned.
• **P 166 a)** Bot. Cent. & top RHS Drawings by Jain. **b)** Bot RHS from "The Secret of the Golden Flower" explained by Richard Wilhelm. • **P 167 a)** LHS Yantra drawn by Prof. Callum Coats on computer. Thanks for the many days & nights you spent pioneering inventive ways of expressing & creating M. Sqs. **b)** RHS Drawing by Jarrah Tree, 1995 Byron Bay (except for collaged symbols)
c) The central symbol on forehead is the Tetragrammaton or the 4-lettered Name of God arranged as a Tetractys within the inverted human heart = Gods Throne, from Böhme's "Libri Apologetici".
• **P 168 a)** RHS Yogi from "Science of Soul" by Vyas Dev shows Astral Body in Brahmacandhra : Causal Body in Heart. **b)** Aura from "The 72 Names of God" from Athanasius Kircher's "Aedipic. Aegyptiacus. This rare Qabalahistic image shows the 72 Names of God in 72 different languages inscribed upon the petals of a symbolic sunflower.
c) Cent. RHS collage on Heart Chakra is the same image referenced for P160 a).
• **P 169** RHS Drawing by Jain, embellished by Sabila Robb 1992.
• **P 170** RHS Drawing by Jain.

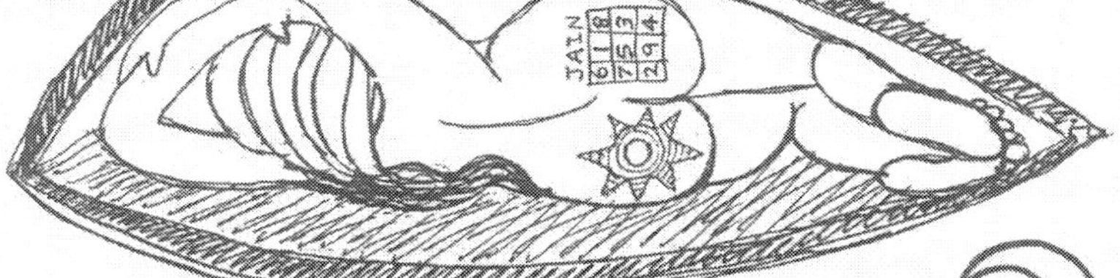

• On this Page a) Top RHS from "Movement of Spiritual Inner Awareness" John Roger. b) Bot RHS. Drawing by Jain c) Bot. LHS Photo by Ryszard Horowitz for AC & R Advertising Inc.

ACKNOWLEDGEMENTS : CHAPTER "X"

• P 172 The image of the "X" of 1 at 4×45° (Atomic structure of Platinum crystal, Fig 160 by Jain) has been tiled 20 times to form a unique border. This is a computer graphic faxed to me by Mary Van de Clark of Traverse City, Michigan. She, as a teacher, is keen to introduce this sacred geometry to the American schools. Has moved address. I dreamt that this image + subject has been released to schools under her name without any due acknowledgement of my genius in "re-discovering" these designs.

• P 174 Credit is given to Naomi of Elands, N.S.W who first presented to me Fig 149. From it, I developed the visual "X" of 1 (Fig 149 a) & then for all the other numbers. Then I realized it had all been done in the Islamic mosque, temple art eg: the Alhambra as depicted in a large book called "Shelter" which showed "A Net of Sixes Overlaid" derived from this "X" Table. No author. Pre 1976.

• P 175 a) Top LHS - a consecrated Magical Sword with Holy Names written in holy hours etc from Eliphas Levi's "The Magical Ritual" b) Bot LHS Drawing by Jain.

• P 179 a) Top Cent. Fold-Out. Byrant Tuckerman was one of the 4 pioneers in the art of Folding Flexagons. Fold down along the solid lines; fold in the opposite directions along the broken lines. Glue edges A & A, B & B etc. From Martin Gardner's "Mathematical Puzzles & Diversions" b) Top LHS the 4 Octahedrons show some of the 9 different planes of symmetry. From Brian Bolts "More Mathematical Activities" c) Top RHS Cent. Hexadekahedroid, from one of Claude Bragdon's books. d) Top far RHS "Diamond Composition with Grey Lines" by Piet Mondrian 1918. e) Cent. The square being hit by the sledge is a M.Sq. of 11×11 by Jain, originally a 3m × 2m cotton banner that is now a dancing robe. f) Top RHS Pyramid from Frontispiece, Hogarth's "The Analysis of Beauty" LSU Middleton Library. g) Bot LHS 2 bodies in diamond symbol for Koh-e-nor University.

• P 180 a) Bot. mid Quote by George Polya. b) Cent. Maltese Cross, logo for State Library of Qld, Brisbane.

• P 182 a) Bot LHS Latin Phrase: "Cogito, ergo sum" by Descartés (French Philosopher): "I think, Therefore I Am". The whole picture has Rosicrucian symbols. Artist unknown. b) Top LHS notice that the white lines in this border are a series of interconnecting swaztikas. This is relevant, as I discovered that M.Sqs. of 6 when rotated 4 times, produce 卍. See Vol 1, my former book. c) Top RHS is the same "X" of 1 at 16 × 22.5°. d) Bot RHS snake from "Transcendental Magic" by Eliphas Levi e) Top LHS & Cent. 2 Drawings by Jain.

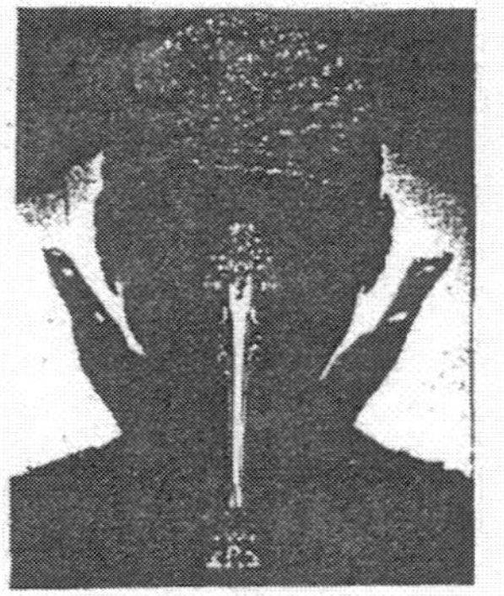

Page 251

• On This Page : Bot. LHS Drawing by Jain
b) Cent. Drawing by Jain
b) RHS Drawing by Jain
• P 183 a) Bot. Cent. Detail from Koran illumination.
• P 185 a) Bot RHS Quote from "The Life & Teachings of the Masters of the Far East" by B.T. Spalding Vol 2 p 56. A highly recommended 5 Volume set.
• P 186 Bot LHS 2 diagram + Bot RHS diagram from "Shelter". See also Acknowledgement notes for P 174.
• P 188 LHS Sphere Spiral from "The Graphic Work" by M.C. Escher.
• P 189 Bot LHS from "The Celtic Art Source Book" by Courtney Davis.
• P 192 John of Taree C.Y.S.S (Community Youth Support Scheme) who gave me permission to muralize a front brick wall at entrance. 2 weeks to paint, Nov 1985. Dedicated to Nikki of Forster, who shortly after the completion of the mural, suicided off the Ellenborough Waterfall, Elands: largest single drop of waterfall in the Southern Hemisphere.
• P 194 a) Top LHS : known as Triton. (½ man & ½ fish) & bears the trident. b) Thai monk in meditation posture. c) RHS words by Jain (a.k.a Haha-Aha [Higher Self]) • P196 LHS drawings + RHS drawings + poetry in circle on Bot. LHS, by Jain.

ACKNOWLEDGEMENTS : SOLUTIONS : CHAPTER ZERO

Page Numbering : Apology

You may have noticed 2 sets of Page Numberings; one originally on the top or bottom Left of each page, when both Vols (2+3) were combined as one book, and a new set of numbers at the top or bottom midpoint of each page, beginning from page 1 and ending at page 174. This was required after Volume 3 was separated from Volume 2.

• P 219 LHS Crucifixion drawing by Jain. In 1982 I lived in Torres Straits on 5 different islands painting murals in their church halls, of Jesus. The aurae of Jesus were elaborate M.Sq. Yantras, but I have no record of them. I thank those 5 chairmen who received me as they would their own son.

This Giant Aperiodic Crystal is the base sequence for the chromosome of bacteriophage φX174. It is the first complete genome ever mapped out for any organism. About 2,000 of these boustrophedonic pages would be needed to show the base sequence of a single E. Coli cell, and about one million pages to show the base sequence of the DNA of a single human cell. The book now in your hands contains roughly the same amount of information as a molecular blueprint for one measly E. Coli cell.

• P220 a) RHS the background aura of this model cut out from a page of DNA mapping, viz: ⟵ b) Bot LHS – the star on her navel is a pattern isolated from the centre of a m.sq. Yantram.
• P 221 RHS Drawing by Jain.
• P222 a) Top LHS "Parabrahm" frontpiece from Franz Hartmann "Magic: White & Black" 1886 reproduced in Le Lotus I (March 1887). b) Bot. RHS "Magic Cross" & Bot. LHS "Six-point Star" & Top RHS "Magic Hexogon" from "Mind-Benders" by Ivan Moscovich. Penguin pub.

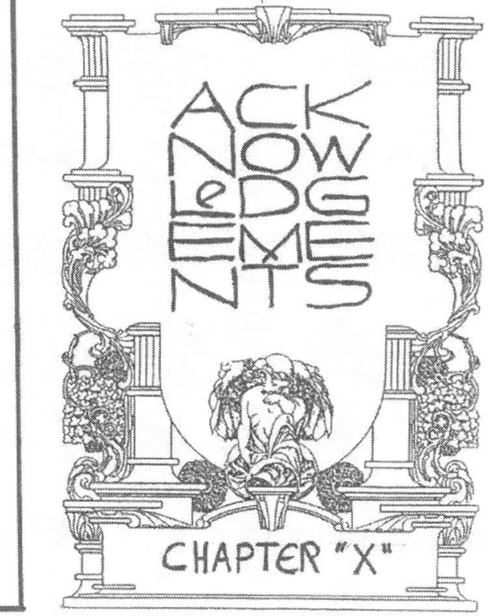

• **On this Page a)** RHS Top & Bot. Universal Energy Field in shape of dough-nut or TORUS. The essential geometry of magnets, trees, galaxies etc
b) LHS metatron's Cube from the Drunvalo Melchizedek material written in Bob Frissell's book "Nothing In This Book is True, But Its Exactly how things Are". **c)** Bot LHS "Fruit of Life" from the same source.
• **P223** LHS Drawing by Jain • **P224** LHS by Bob Gill: "Forget All The Rules" book jacket 1981
• **P225** LHS by H. Armstrong Roberts 1988 Card. Netherlands.
• **P227** 2 Drawings by Jain • **P229** 2 Drawings on LHS & RHS by Jain, both are long continuous unbroken lines. Fun to do.
• **P230** LHS "Applicants For A Sou" Mentonese children, from a painting by James Sant, R.A. exhibited in 1869 Royal Academy.
• **P231 a)** LHS Drawing by Jain.
b) The m. Sq. of 3 at the base of this drawing is written in Sanskrit. Shown here is its English translation. Sometimes the sacred central cell of "5" is left blank, to denote the Void or the never-ending Golden mean Spiral that is generated by the ★. Five = phi-ve (φ or Phi denotes the mathematics of Nature, atoms, flowers, etc.

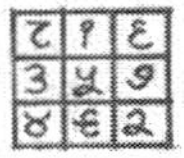

→

8	1	6
3		7
4	9	2

• **P232 a)** Bot LHS Drawing by Jain.
b) Bot. Cent. is a Computer-Produktion in den. USA Umfassender AusleseprazeB. The collaging of m. Sqs with top computer buisnesses in this world is a prayer to humanity to utilize this technology as a service to the fullest Expression of the Light. May all Companies Love.
• **P233** Top LHS Drawing by Jain.
• **P236** Bot LHS Drawing by Jain. • **P305** LHS Brocolli-like Pythagoras Tree collaged 4 times around a square centre. Construction with isosceles triangles which have angles greater than 90°. From "Chaos + Fractals" by Peitgen/Jürgens, chapter: "Classical Fractals & Self-Similarity".

I Pray that these spiritual mathemagical Geometries be knitted, embedded & traced 3-Dimensionally into your Aura to begin the true healing process known as

REMEMBERING
of The GOD I Am
the Essence of Love You Are
The Beginning

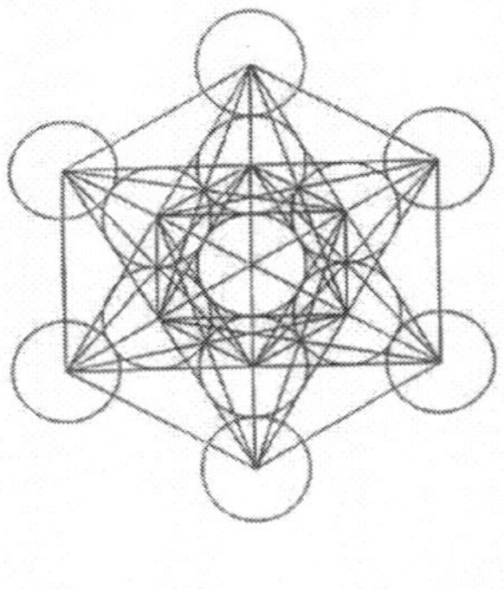

"WE ARE NOT HUMAN BEINGS HAVING A SPIRITUAL EXPERIENCE WE ARE SPIRITUAL BEINGS HAVING A HUMAN EXPERIENCE".

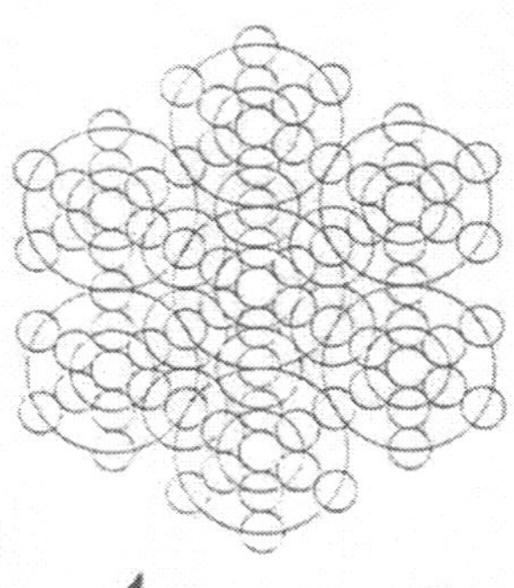

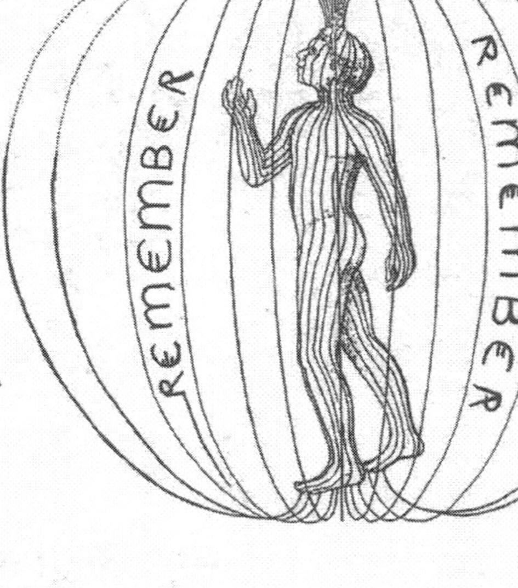

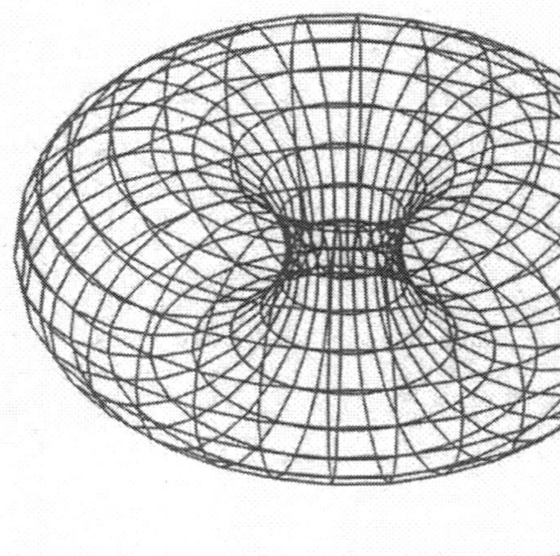

THE END

PAGE 253

ACKNOWLEDGEMENTS : DID YOU KNOW ?

PAGE 237 : The 3 Squares/Cubes on the RHS are from "The Cube of Space" Container of Creation by Kevin Townley 1993 published by Archive Press.
PAGE 238 : Bot. RHS The Light-Body Network from "The Crystal Connection" by R + V Baer 1986. **PAGE 239 :** Bot. RHS Drawing by Jain depicting Jennifer Sewell's Angel Guide in 1997. Top + Bot. LHS + top RHS taken from "Sufi" by Laleh Bakhtiar.
PAGE 240 : Top RHS Diamond Lattice taken from "Extra-Sensory Perception of Quarks" by Stephen Phillips 1946.
PAGE 241 : Top + Bot. RHS Diabolic Doughnut + HyperCube taken from "More Mathematical Puzzles + Diversions" by Martin Gardner. Top + Bot. LHS taken from "The Magic Square of 3 Crystal" by Arto Heino.
PAGE 242 : Top RHS Egyptian M. Sq. taken from "Magic: An Occult Primer" by D. Conway. Bot. RHS Mason's Mark M. Sq. 3 taken from "The Sufis" by Idries Shah. **PAGE 243 :** Bot. RHS Hexagonal DiPyramid with M. Sq. of 3 nos. + adjacent hexagon taken from Arto Heino's "The Magic Square of 3 Crystal".
PAGE 245 : RHS. A Sino-Tibetan Zodiac, Trigrams and Magic Square of 3×3, based on a Tibetan block-print. Taken from: "The Inner Structure of the I-Ching" by Lama Anagarika Govinda.

ACKNOWLEDGEMENTS for OTHER Graphics used in Volume 3:

Inside of the Front Cover: Painting of a Hathor by Jain 1999. Hathor is a cow-eared Egyptian Goddess that resides in the Higher Dimensions (4th) of Venus. As a Spiritual Tribe, the Hathors make visits to us invisibly and may tickle our ears with feathers to make us laugh when we get too serious, pedantic + overwhelmed by life. This was painted + gifted to Leigh Shorthouse during her 7½ months pregnancy for the spiritual protection of our son.

PAGE 255 : LHS. "Infinity" graphic "Fractals, Chaos, Power Laws" by Manfred Schroeder. published by W.H. Freeman + Co. 1991.

The Magic Square of 3×3 "LO-SHU" is the Centre of the Tibetan Calendar. The magic Sum or CONSTANT is $C=\frac{n}{2}(1+n^2)$

APPENDIX
NEXT PAGE

BIBLIOGRAPHY

Logarithmic spiral to infinity.

1 – "The Gospel of The Holy Twelve". translated from the original Aramaic. 1974.
2 – "Magic Squares & Cubes". W.S. Andrews. Dover. 1917.
3 – "Islamic Patterns". Keith Critchlow. Thames & Hudson. 1976.
4 – "Games Ancient & Oriental". Edward Falkener. Dover. 1892.
5 – "Mathematical Recreations". M. Kraitchik. Dover. 1942.
6 – "Mathematical Recreations & Essays" W.W. Rouse Ball. Macmillan. 1892.
7 – "Mathemagic". R.V. Heath. Dover. 1933.
8 – "New Recreations with Magic Squares". W.H. Benson & O. Jacoby. Dover 1976.
9 – "Mandala". José & Miriam Arguelles. Shambhala. 1972.
10 – "Earth Ascending". José Arguelles.
11 – "Mayan Factor". José Arguelles. Bear & Co. 1987.
12 – "Islamic Science". Seyyed Hossein Nasr. World of Islam Fest. Pub. Co. 1976.
13 – "Sacred Geometry" Robert Lawlor. Thames & Hudson. 1982.
14 – "Encyclopaedia Brittanica". 1970.
15 – "Solutio" containing Bernard du Bessy Frenicle's 880 Basic Magic Sqs. of 4×4. 1605-1675. Indexed & inventoried by de Haas & Sijn.
16 – "Magic Squares". John Lee Fults. Open Court. 1974.
17 – "More Mathematical Puzzles & Diversions". Martin Gardner. Penguin. 1961.
18 – "Mathematics". Life Science Library. David Bergamini. Time Inc. 1965.
19 – "Magic Square Puzzles" David King. Frederick Muller Limited, London. '84.
20 – "Number Play". Gyles Brandreth. Severn House Publishers, London. 1984.
21 – "Mathematische Mußestunden". Prof. Dr. Hermann Schubert. Berlin. 1941.
22 – "Optical & Geometrical Patterns & Designs". Spyros Horemis. Dover. N.Y. 1970.
23 – "Projective Ornament". Claude Bragdon.
24 – "Beyond Pyramid Power". G. Pat. Flanagan. 1975.

Continued Next Page.

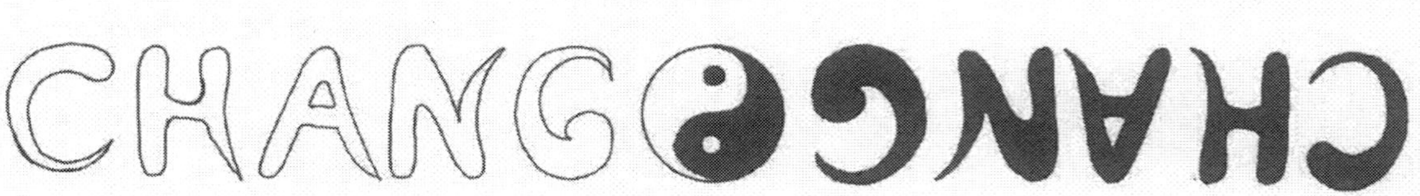

Graphic, from "Fractals, Chaos, Power Laws" by Manfred Schroeder. W.H. Freeman & Co. 1991.

Bibliography Continued.

Lo-Shu to Thee I raise my whole being,
—a vessel emptied of self. Accept, Lo-Shu,
this my emptiness, and so fill me with
Thyself—Thy Light, Thy Love, Thy
Life—that these Thy precious Gifts
may radiate through me and over-
flow the chalice of my heart into
the hearts of all with whom I
come in contact this day
revealing unto them
the beauty of
Thy Joy
and
Wholeness
and
the
serenity
of Thy Peace
which nothing can destroy.

—Prayer and calligraphy by Frances Nuttall—

THE PRAYER OF THE LO-SHU

25 — "Practicals of Yantra". L.R. Chawdhri. Sagar Publications 1984.
26 — "Tibetan Tantric Charms & Amulets". Nik Douglas. Dover N.Y. 1978.
27 — "Yantra". Madhu Khanna. Thames & Hudson.
28 — "Science of Soul" (Atma-Vijnana) Shree Swami Vyas Dev Ji Maharaj. '64.
29 — "Yantra Vidya" Siddha Baba Hoddarnath. 1986.
30 — "Sri Chakra". S.K. Ramachandra Rao. Sri Satguru Pub'n. 1989.
31 — "I-Ching". The Chinese Book of Changes. p310 Wilhelm's translation. 1968.
32 — "The Trigrams of Han". Steve Moore. Aquarian Press. 1989.
33 — "Shelter" edited by Lloyd Kahn.
34 — "Planet Heartworks". Dan Winter. 1988.
35 — "Sacred Geometry" Video by Dan Winter of Crystal Hill Farm.
36 — "The Magus". Francis Barrett. Citadel Press. 1801
37 — "The Books of the Beast" T.D. Smith. Crucible. 1987.
38 — "Alchemy Today" Vols 1 & 2. Chris Illert. Science-Art Library 1992.
39 — "The Secret Teachings of all Ages". Manly. P. Hall. 1928.
40 — "City of Revelation" John Michell. Abacus. 1972.
41 — "The Keys of Enoch" J.J. Hurtak. Academy for Future Science. '75.
42 — "Nothing In This Book is True..." Bob Frissell. Frog Ltd. 1994.
43 — "How To Make & Use Talismans". Israel Regardie. Aquarian Press. '72.
44 — "By The Holy Tetraktys". L. Gordon Plummer. Point Loma Pub'ns. 1982.
45 — "Timaeus". Plato. Penguin.
46 — "Life of Pythagoras". Iamblichus. Translated by Thomas Taylor.
47 — "The Inner Structure of the I-Ching". Lama Anagarika Govinda. 1981.
48 — "Robert Fludd" material circa Cth16. Jocelyn Godwin. Shambhala 1979.
49 — "Spiral Symmetry" I. Hargittai & C.A. Pickover.
50 — The Buckminster Fuller material e.g: "Utopia or Oblivion". Pelican. 1969.
51 — "The Divine Proportion". H.E. Huntley. Dover. 1970.
52 — "The Power of Limits" (Proportional Harmonies in Nature, Art & Architecture). György Doczi. Shambhala. 1981.
53 — "Harmonic Mathematics". (A Phi-Ratioed Universe As Seen through Number Harmonics). W.B. Conner. 1982.
54 — "The Cunning Caliph". Kenneth Kelsey. Frederick Muller Ltd, London. 1979.
55 — "Encyclopedic Dictionary of Yoga" Georg Feuerstein. Paragon House. 1990.
56 — "Encyclopedic World Dictionary". Paul Hamlyn Publishing Group. 1971.
57 — "Mathematics In The Making". Lancelot Hogben. Galahad Books.
58 — "Rhythms of Vision". Lawrence Blair. Shocken Books. 1976.

Albrecht Dürer's 'Melencolia'
At the upper right is a Magic Square

JAIN's BIOgraphy :

1957 – Born 13th Oct. Sydney. Sunrise. Named "Collin Nicholas Saad". Both parents emigrated from the Lebanon mountains, their lives based on subsistence agriculture. Learnt building / bricklaying skills from my father Peter Saad, since the age of 5, assisting him as he climbed from rags to riches. Mother: Yvonne Khoury, whose father was a village priest.

1969 – Age 12, Went with family to New Zealand on a Russian cruiser "Shota Rustavelli" that had a giant 4m x 4m Magic Sq. of 3x3 painted on its deck

1972 – Age 15 travelled to Lebanon for 4 months with family. Met a 10 yr young girl who presented to me a M.Sq. card trick of 9 cards to sum up to 15. She never gave me the answer & thus planted a seed of curiousity + research.

1974 – Age 17 was selected as one of the top mathematicians to attend studies at a nuclear reactor, Lucas Heights, west of Sydney.

1975 – Declined a medical career to study 2 yrs of Professional writing + Psychology at the Institute of Technology, Sydney. This created a major dissension with my father + I. Self-taught Astrology + Tarot + Gematria.

1979 – Age 22. Travelled again with parents to France, Italy, Greece & mainly Lebanon for 4 months. Interested in Temple Building + Magic Squares.

1981 – Age 24. Printed my first book of poems + art + prayers + geometries: "The Time Twins" by Enfant-de Neptunuu. Sold out immediately.

1981
1982 – Age 24, 25. studied herbal medicines with Aboriginal Elders in Torres Straits & islands off New Guinea. Designed a political flag for the chairman of Badu Island. In exchange for living on 5 islands, I was commissioned to paint 4 large 3m x 3m wall murals of Jesus on Church annexes. These were my first M.Sq. Murals, the radiating geometries representing Jesus' aura/chakras.

1983 – Lived for 1 yr on 'Arthaven': a mountain-top community, painting another 3 large murals Blackheath, Blue mtns. These were grafittied & led me to protect my work by painting directly on cotton sheets. Portable Art that I can exhibit anywhere & travel with.

1984 – Changed my name to JAIN. First major injury. Stabbed thru chest 6" knife wound. An Out-of-Body experience ensued giving me clear purpose of my Path to come back in the same body & teach Sacred Geometry/M.Sqs. to children + write many books like this.

1984
1985
1986 – Many months of fasting ensued, to heal. Hermited for 3 years near Elands, mid-coast N.S.W. where I painted 7 large 3m x 3m cotton tapestries of the M.Sqs. Also prepared thousands of A3 sized pages / manuscripts for future publications. In '86 I travelled to Ayers Rock, for 6 months on the "Peace Train", teaching M.Sqs. in schools for the first time eg: Apollo Bay High School + Adelaide Hills.

1987 – Met Reenah Sun, who lived with me for 7 yrs, having 2 children with her: Mingkah Jain Sun & Asha Jain Sun. cont...

BIO continued

1988 – Moved to Byron Bay hinterland area painting more 3m x 3m M.Sq. mandala banners & teaching sacred geometry + meditation in schools.

1989 – Travelled with Reenah + Mingkah (age 2) to Kashmir + Ladakh (Tibet) for 4 months. Met a Magic Square Yogi + Dalai Lama's astrologer who transmitted much knowledge to me plus permission to disseminate these yantras worldwide.

1990 – Travelled to Thailand for 4 months studying comparative religions + engaging in Vipassana Meditations (10 days silence). On return I manifested $10,000 & printed my 2nd book "The Book of Harmony Squares" Vol 1. Sold well to many schools & universities. Asha, my second daughter born.

1991 – Made a series of mandala table-tops of broken-tiled mosaics.

1992 – Created & silkscreened 3 designs of M.Sq. Decals (adhesive transparencies) for the International market.

1993 – 4th Exhibition of my M.Sq. Tapestries, in Byron Bay. Sold a major painting of a M.Sq. of 7x7 superimposed with a Jesus figure, for $3,000. I lectured there to 40 mathematicians. Event videoed. Taught also at Cairns schools for 1 month. Also created the first M.Sq. Computer Program with Anthony McCardwell of Paradise Valley.

1994 – Lived in Tallebudgera Valley for 1yr where I created, by needle + thread, the first Human Cloth Book that holds 30 large cotton pages 1200x1200 mm containing the essential M.Sq. Knowledge. Basically to teach 'Maths' via medium of theatre.

– Indepth library research of M.Sq. manuscripts centuries old from overseas: Inter-Library Loans eg: accessing fotocopied records of Frenicle's 880 permutations of the M.Sq. of 4x4 from C16th. Creation of a rare Library.

– Invited to teach, in Qld, 140 dux-of-the-schools teenagers at a Maths Camp, as a Prof. of Magic Squareology.

– First performance of "The Theatre of the Holy Numbers", a M.Sq. based ritual theatre, interactive with the audience, the concept being to entertain as well as leaving the audience with Knowledge of Spiritual Maths.

1994 – Received a letter from Chris Illert, a nuclear Scientist verifying that my M.Sq. formula $C = \frac{n}{2}(1+n^2)$ the Constant or Magic Sums relates to the atomic onion-shelling of electrons. As a mystic, 15 yrs ago, I intuitively knew this.

1995 – Invited to teach M.Sqs theatrically at a Gathering of 100 mathematicians at the Southern Cross Math's Association, Lismore. Maths President requests a CD-Rom made.

– Invited to teach at a 2nd Maths Camp, Lennox Heads, Nth NSW for 14 yr olds.

– Invited to exhibit my Life's work at the Brisbane "Health, Body & Harmony" Trade Fair. Very successful. Met Ajay of New Dimensions who now is my Distributor.

– I co-facilitated a Sacred Geometry 4 Day Workshop with Karen MacDonald, on the MerKaBah as taught by Drunvalo Melchizedek: Flower-of-Life seminar Founder.

1996 – Invited to teach M.Sqs. to another sacred Geometry group in Brisbane with K. MacDonald.

– Pass exams at Bangalow N.S.W giving me legal Accreditation to teach at Primary schools the Arts of M.Sqs, Meditation + Sufi Dance.

– Organized 30 friends to perform "The Theatre of the Holy Numbers" with a well rehearsed Harmonic Choir (Rodrigo + friends) for an Equinox production in Byron Bay with an audience of 500 people. Event videoed & advertised on Internet.

– Received verification that Prof. Searl's life's work is based on the Mathematics of constructing M.Sqs! His books are called "Law of the Squares". In 1971 John Searl created levity discs ie. anti-gravity UFO-like machines, free energy generators that can power cars, houses, for free; clean & non-polluting! Thus I teach that Mathematics & Magic Squares are essential to understanding the Science of Free Universal Energy & the realm of the Nuclear Platonic Geometries.

1996 – Ready to print my 3rd book "The Book of Magic Squares" Vol 2, having taken 5 years to prepare the manuscripts. Expected to be released at end of 1996.

THE COSMIC GLOSSARY

LEGEND

- M. Sq. = Magic Square
- Ab. = Arabic
- Gk. = Greek
- Heb. = Hebrew
- Lat. = Latin
- Skt. = Sanskrit
- lit. = literally
- sing. = singular
- a.k.a. = also known as
- K.O.E. = The Keys of Enoch
- "CAPITALS" = that word is listed in this Glossary.
- This Glossary also contains listings or explanations of words from "The Book of Harmony Squares" Vol 1. 1990.
- ∴ = therefore.
- pp = Pages
- p = Page
- No. = Number

ABRACADABRA
ABRACADABR
ABRACADAB
ABRACADA
ABRACAD
ABRACA
ABRAC
ABRA
ABR
AB
A

ABRACADABRA : Heb. "Ha-Brachab-Dabarah" lit = Speak or pronounce the Blessing, to invoke the Holy Name of Jehovah (YHWH). Usually inscribed inside an inverted triangle. This word was in frequent use during the Middle Ages as a magic formula relating to Abraxas of the Gnostics, or one of the names of the sun-god Mithras (Persian: the God of Light & Truth, later of the Sun, a major source of X'ian belief).

ARCARYA : Skt. lit. 'preceptor' who teaches to a pupil via the word of mouth. A spiritual master who teaches by example. The Mahābhārata epic (XII.313.23) likens the preceptor to a ferryman & his knowledge to a ferry.

ACTA-KALAPAS : Pali, lit. 8 Subatomic particles reflecting 8ness or (Acta) as revealed by Gautama the Buddha during deep meditational insight (VIPASSANA). Siddha Yogis can enter the Microcosmic realms.

ADAM KADMON : Ab. the primordial man who is the reflection of the Divine. p52 shows the Qabalahistic divine image in prayer. See "QABALAH" in glossary.

ADONAI : Heb. lit: 'Lord'. Any manifestation of higher Light Consciousness or manifestation of a Lord representing the Divine Mind (K.O.E.). Adonai sums to 34 & thus relates to the secret code Name for the M Sq. of 4 x 4 whose CONSTANT or magic sums = 34. In the Latin Vulgate, Adonai is translated as "Domine".

AH : Skt. The first vowel sound of the sanskrit alphabet. A tantric sound syllable signifying the beginning of all Creation. The Sound of deep lov-making. See also "HAHA-AHA".

AHA = the sound of Divine Knowing or Realization of Remembrance. The Vocalized Sound of Intuition. Tibetan universal mantra.

AHAM BRAHMA : Skt. lit. I AM God. God I AM.

AHIMSĀ: Skt. "non-harming" or non-violence physically, mentally or orally at all times.

AIN: Ab + Heb, also 'Ayn' = Eye, spring water, a well. It is the Absolute & Limitless (Void), the No-Thingness. "The Eye is the Light of the Body; if ∴ thine Eye be single, thy whole body shall be full of Light" (Bible, Matthew). AIN SOPH. Heb. The Limitless Light, the Light that "sees" & "knows". Also, "AIN SOPH AUR". (K.O.E).

AJNA: Skt. The CHAKRA or spiral of psychic energy which activates the brow center in the subcortical areas of the frontal lobes. 3rd Eye. The 6th of the 7 Chakras, governs our intuition + abstract knowledge. See "CHAKRA" + p 157.

AKASHA: Skt. lit. "radiance", located in the heart, equal to 100,000 suns. Symbolic of the transcendental Self's ether/space. The realm that holds the World Records. The Ultimate Vibrational Library of every thought, emotion or event of All Time. AKASHA - CHAKRA = ether wheel. 9th psycho-energetic center in the human body.

AL-ARQĀM AL-HINDIYYAH: Ab. lit. Indian Numbers.

ALCHEMY: Prescientific craft of using natural elements to produce physical gold & immortality. The quest for a divine body. "Al" = 'the' & "Kimia" in Egyptian = secret. "Khem" was also the ancient name for Egypt. They produced a white-powdered Gold (of Iridium, etc) for healing; a panacea for the masses as prophesized by Nostradamus.

AL-INSAN AL-KĀMIL: Ab. Universal Man, in whom the fullness of the human state is realized & through whom multiplicity returns to Unity.

AL-JABR: Ab. Algebra, originally meant restoration & amplification of something incomplete.

ALPHA & OMEGA ($\alpha + \Omega$ or Ꭿ): Gk. lit. 'first & last' letters of the Gk Alphabet. They represent the joining of No.1 to the last No. of the Magic Square. This FINAL CONNECTING LINE, when viewing the YANTRA as a harmonic electrical circuit board of copper wires, actually closes the circuit in such a way as to create a flow of foreverness. Eternal Free Energy. Read "SEARL (Prof)", "BALANCE OF THE PAIRS", "COMPLEMENT" & "RODLIN COIL DESIGN".

AL-'ULŪM AL-NAQLIYYAH: Ab. the transmitted sciences; knowledge acquired through the path of revealed truth which after its revelation is transmitted from generation to the next.

AMAROLI: Skt. The Immortal Nectar. The Yoga-Tattva-Upanishad (128) describes it as the daily drinking of the amari, or urine. Drinking only the middle flow of one's urine or Shivambu (Water of Shiva). MAHATMA Gandhi (1869-1948) drank it religiously all his adult life. Cures all ailments including cancer. See also "AMRITA".

AMBAJI: Skt. One of the 9 divine spouses of Shiva God. She rides a tiger & carries fierce weapons in her 8 arms that conquor ego, lust, greed etc. As a Goddess of Destruction she is a benign loving force to her devotees. Her tantric symbol is the magic Triangle whose summations of the 3 sides all equal 20, 20, 20. The 9 numbers represent the 9 aspects of the Goddess Shakti, Kali, Durga, Chamunda etc.

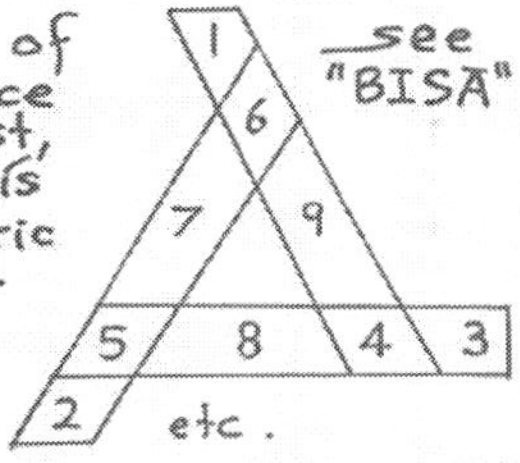

see "BISA"

AMEN: Gk. lit. "verily, truly". Heb. means "strength" stability, so be it. To say "amen" to, means to be in agreement with. Uttered at the end of a prayer. The Skt. version is "Aum" or "Om".

AMRITA: Skt. lit. "immortal". It refers to the nectar of immortality that trickles down from an esoteric center in the head. This ambrosia produces a superior body endowed with enormous strength & vigor, disease free & agelessness.

ANĀHATA-CHAKRA: Skt. lit. "Wheel of the unstruck sound". Heart Centre, the seat of the Divine. See CHAKRA + p 157.
ANAHATA GURU = my Heart is my Teacher. see p 113.

ANĀL HAQQ: Ab. lit. "I Am The Truth".

ANANDA: The ultimate state of consciousness in which nothing is experienced except Blissful at-one-ness with all things great & small. The simplicity of Love & consciousness of the One. We are born in bliss & ∴ every human has experienced Ananda.

ARCHETYPE : Gk. "Archetypos". lit. original, first-moulded. ∴ model or first form. The original model or pattern after which a thing is made. In Magic Squareology, these codes exist in all cultures & is moulded into the mass-consciousness & cellular memory. We know Everything, we've only Forgotten. p 188.

ARDHAMATRA : Skt. the mathematical descent from the BINDU (Centerpoint, Sperm, Drop) to the unified & ordered periphery. ("rdh" = to grow & "matra" = measure). see p 95 + BINDU.

ARDHANAGESHWARI : Skt. SIVA-SAKTI as one embodiment, as pictured on p 101. It reflects our divine Androgyny !

ARITHMĀTIQI : Gk. a term for the science of numbers derived from the ancient Greek culture.

ARYAN : Skt. one whose life is based on spiritual values.

ASANA : Skt. A physical posture & exercise to relax & refine the tensions in the body so that the mind does not pay attention to it & the intellectual hang-ups do not get in the way of the free flow of the psychic processes. Releases ancient traumas in the cells & bones.

ASCENDED MASTERS : Masters who have served several incarnations in the lower heavens teaching the Cosmic Law of the Universe & who have ascended back into the presence of the All-That-Is from whence they receive new assignments to teach a wide variety of worlds because of their greater love. (K.O.E).

ĀSHRAMA : Skt. from "shrama" lit "effort". A hermitage where an adept instructs disciples who exert themselves in a sacred way of life. Most are BRAHMACARYA i.e. celibate.

ASHINS : Skt. ½ human-bodied & ½ horse-headed beings who reside in the devic realms. Physicians to the Gods of BHARAT. p 32.

AUXILLARY SQUARES : or Imaginary Squares; are the same size as the main square that will become a M.Sq. that surround the main square, collect moving numbers & are lifted up, or superimposed upon the main template square to construct & create a final M.Sq. see p 73 fig 52.

ASMI : Skt. lit. "I AM". "God I Am". "I Am Source". I Am a source of Violet Fire greater than any human Experience.

ASSOCIATED : When the complementary pairs or diametrically opposite Nos. of a M.Sq. are equidistant from the center. see p 67 fig 47, p 85.

ĀTMĀ : Skt. the eternal Spirit soul who is living within the temporal, material body.
ATMA-JNANA : lit. Self Knowledge. The form of the Self that cannot be seen by the eye. The Self reveals itself only to itSelf.
ATMAN : Pure consciousness of the undivided soul which is a spark of the Divine nature. The unmanifested essence of the ultimate Self which programs the creative forces with its own limitations & self-sense. The real "I" who is at the back of the phenomenal, mental & emotional "I".

AVARANA : Skt. One of the 9 tiers of the SHRI YANTRA as an abode of the Goddess. All together they are known as: Navavarana ("Nava" = 9). Visualized 3-Dimensionally, the last tier is the BINDU or jewelled peak of a mountain or palace. see p 98 + MOUNT MERU + BINDU.

AVATĀRA : Skt. One who descends from the spiritual sky. Incarnation or descent of the Supreme Lord. e.g: BabaJi.

AWQĀF : Ab. religious endowments, plural of "waqf", relating to institutions such as schools, libraries & other educational establishments eg: "F. R. E. E. D. O. M. S.".

Ya AZIM : Ab. "O Most High". A Darvish (SUFI) greeting meaning: "How beautifully does God manifest through you".

BALANCE OF THE PAIRS : the connection of first & last number of a M.Sq. e.g: for any MAGNITUDE let 'n' = m.sq., then the formula is "1 + n²". This de[mo]nstrates more internal harmonies & is part of the spiritual mathematics that is a clue to the nuclear geometries. e.g. p 19. see also "ALPHA + OMEGA" + "COMPLEMENT".

BHAGAVAD-GITA : Skt. lit. Lord's Song. Krishna's LILA.

BHAKTA : Skt. To discover the Divine in all beings & in all forms. To acknowledGe The GOD•DESS•ESSENCE.

BHAKTI YOGA : Skt. lit. "Yoga of Devotion". A devotee of God, utterly devoid of personal motivation apart from the desire to become God's instrument.

BHARAT : Skt. the ancient original name for India.

BHAVA PRATYAYA : Skt. Born with extraordinary powers.

BHUPURA : Skt. An enclosure of triple lines around the eg: SHRI YANTRA, as spiritual protection. It has portals or gateways (dvāra) which means only the honest & devout can enter into the mysteries of SHRI YANTRA. see p 97

BI-FOCCI : has 2 centers or focci points as demonstrated graphically on p167 fig 143 & p 223 fig 124.

BINDU : Skt. Drop, dot, source point ("bija") or centerpoint of a YANTRA. The origin of all manifestation notably all sound (Nada) as captured by the sacred syllable OM. In TANTRA, Bindu = **Semen**. Loss of semen is detrimental to spiritual growth ∴ sexually active aspirants must learn to circumvent seminal discharge, to become deathless. Vajroli-mudra is the sucking back of semen up the spine to arrest ejaculation This activates KUNDALINI or serpent power.

BISA YANTRA : Skt. "Bisa" = 20. An Indian M.Sq. whose sums or CONSTANT = 20. These sacred ancient codes are found commonly on Indian Goddess posters, like AMBAJI. see p 93 fig 68. These Bis Yantras shown here belong to AMBAJI; has opposite pairs summing to 10; & 4 digited alignments summing 20.

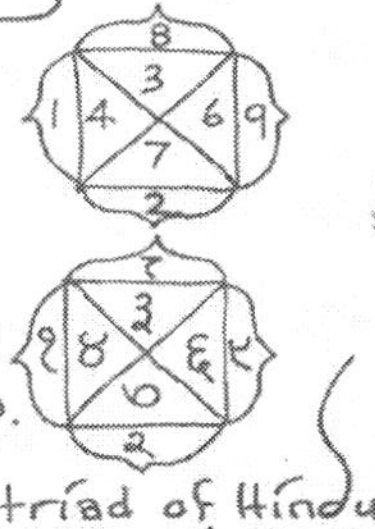

BRAHMA : Skt. Creator God in the classic triad of Hinduism. The other two being VISHNU the Preserver & SHIVA the Destroyer. BRAHMACARYA is the ideal of chastity so as to preserve the great power inherent in the semen (BINDU). This virtue gains great vitality. BRAHMAN is the impersonal aspect of the Absolute Truth.

BREAKMOVES : In the construction of M.Sqs. from NATURAL SQUARES, certain repetitious moves or interruptions are made at every multiple that corresponds to the order or MAGNITUDE of the M.Sq. eG: See P 73 fig 52 + P 75.

BRIOLETTE : Pear-shaped gem with many triangular facets. p 152. Descriptive of the MERKABAH Ascension Vehicle.

BROTHERHOOD OF LIGHT : Advanced spiritual intelligence that can take on physical form & have the responsibility of governing stellar orders with respect to the local hierarchy federation of the Deity. (K.O.E.).

BUDDHA : There are many Buddhas or Awakened Ones, not to be confused with Gautama the Buddha, a Buddhist founder.

CAITANYA : (1486-1533) revived Krishna devotion in eastern India via singing & dancing. A man of extraordinary charisma that during his lifetime he was revered as an incarnation of God Krishna. (lit. "the All Attractive One" cowboy sage who enjoyed a lifetime full of LILA).

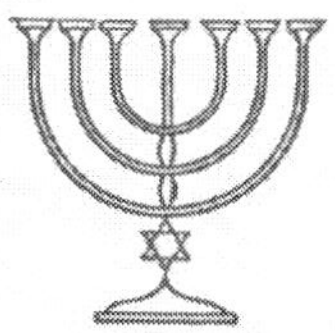

CANDELABRA : symbol of spiritual light & of salvation. The number of its branches has always a cosmic or mystic significance e.g: the Hebraic 7-branched candelabra corresponds to the 7 heavens & the 7 planets.

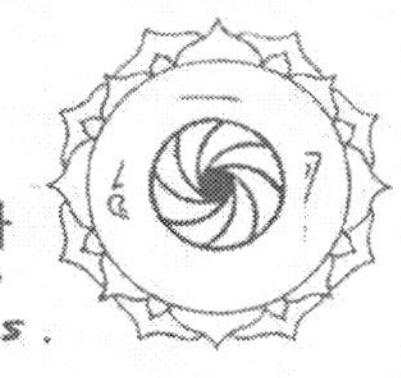

CHAKRA : Skt. lit. "Wheel". Psycho-energetic vortices forming the major organs of the body composed of life energy (PRANA). There are 7 of these lotuses ("padma") + many more. Contemplation upon these centers leads to the conquest of death & to the acquisition of paranormal love powers. see p 120 & next page p 259 diagrams.

CODONS : The 64 DNA Codons are the amino acid "code Words" of the genetic code. These little 6-part structures come together to create the twisting double helix patterns of DNA. The DNA code corresponds to a whole number formula which represents a binary progression to the 6th power : 2, 4, 8, 16, 32, 64 producing 64 six-part code words or codons. Notice how the binary patterns of the Ben. Franklin M.Sqs. of 8 & 16 (on pp 198 & 228 respectively) look like 3-Dimensional spirals of DNA. For more research, see "I - CHING".

COMPLEMENT : referring to the construction of M.Sqs, it is the number that you must add to or subtract from to make the sum of $(1 + n^2)$ which is the sacred sum of the first & last numbers of the M.Sq. see p 27. see "ALPHA + OMEGA" + "the BALANCE of the PAIRS".

THE CHAKRAS OR PSYCHIC ELECTRICITY CENTERS

Psychic Centres	Physiological Counterparts
SAHASRĀRA-PADMA Crown Centre Seed-syllable: 'OM' In the Tibetan System conceived as One Centre (*pdab-ston*) ĀJÑĀ-CAKRA (between the Eyebrows) Seed-Syllable: half or short 'A'	BRAIN Volitional Nerve-system Cerebro-spinal Nerve-system MEDULLA OBLONGATA Non-volitional Nerve-system
VIŚUDDHA-CAKRA Throat Centre Element: 'Ether' as substrate of Sound (*śabda*). Seed-syllable: 'HAM' Colour: White Form: Circle	PLEXUS CERVICUS Respiratory System
ANĀHATA-CAKRA Heart Centre Element: 'Air' (Motion) Seed-syllable: 'YAM' Colour: Grey-Blue Form: Hexagram	PLEXUS CARDIACUS Circulatory System (Blood-vessels)
MANIPŪRA-CAKRA Navel Centre Element: 'Fire' Seed-syllable: 'RAM' Colour: Red Form: Triangle	PLEXUS EPIGASTRICUS (Solar Plexus) System of Nutrition

COMPOSITE: made up of various parts or elements e.g: a Chinese m.Sq. of 9 is made up of 9 Lo-shu's or m.Sqs. of 3. see pp 84–87 or the m.Sq. 12 on p 144 is made up by a knowledge of the m.Sqs. of 3 and 4.

CONSECUTIVE: following one another in uninterrupted succession. Successive. Sequential. Natural Counting order e.g: 1→2→3→4→5→6→7→8→9. see pp 23–25.

CONSTANT: The sum of a magical square's columns, rows & diagonals that are harmonically equal, thus allowing for electronic symmetry. Magic Sum or Line Value. e.g. the Chinese m.Sq. of 6x6 shown here, Bottom R.H. Side from 1590 A.D. has a Constant of 111 or C = 111. These eternal, fixed sums are the secret code names for the many names of Allah.

COUNCIL OF NINE: symbolized by the 9 cells of the m.Sq of 3x3, it represents "the Tribunal of Teachers governing our super-galactic & galactic region, subject to change in evolving 'new programs' of the All-That-Is's Kingdom, such as the regeneration of the lower worlds. (from K.O.E = Keys of Enoch).

CYBERSPACE: Gk. "Cyber" = Pilot. Flowing to the Point where all knowledge collects. Ecologically sustainable development. Self-contained computer systems, like Internet. p 65.

DAKINI:

The Dakini, symbolical of Shakti in its spontaneous intuitive form, represents the guiding and unifying force of inspiration and grace. These Dakinis are the genii of meditation and the spiritual helpers who keep the sadhaka from illusion of the worldly life. They can be embodied in human form unconsciously or in Divine form as visions, promptings, messengers or magic door-openers. A guardian Angel is the nearest Western conception of a Dakini although this does not explain that they are not "Beings" existing outside of ourselves but spiritual impulses and forces which can take any form and manifest any appearances from the unconscious. To a scientist a Dakini could appear as a Divine Computer.

DEDICATION: This book is dedicated to our Forefathers & Five Mothers.

DEVA: Skt "shining one", like Shiva or Kali. Hindu Gods or Christian Angels that exist in the heavenly realms a.k.a. "Loka" = Planet.

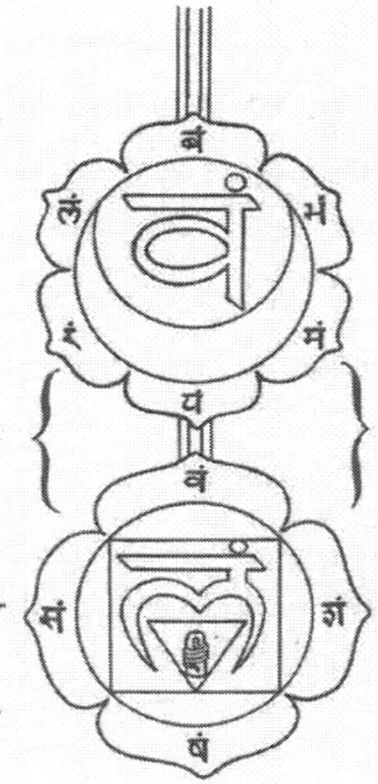

Psychic Centres	Physiological Counterparts
SVĀDHIṢṬHĀNA-CAKRA Abdominal Centre (4 finger-widths below the navel) Element: 'Water' Seed-syllable: 'VAM' Colour: White Form: Crescent	PLEXUS HYPOGASTRICUS Inner Organs of Secretion and Reproduction
In the Tibetan System combined under the name 'Sang-Nā' (*gsaṅ-gnas*)	Generative System
MŪLĀDHĀRA-CAKRA Root Centre (in the perineum). Its latent primordial force is represented by the serpent 'Kuṇḍalini', coiled round the 'Liṅgam' in the centre of the triangular 'Yoni'. Element: 'Earth' Seed-syllable: 'LAM' Colour: Yellow Form: Square	PLEXUS PELVIS (Sacral Plexus) which controls the outer organs of generation (represented by 'Liṅgam', the male, and 'Yoni', the female symbol of creative force (comparable to the 'libido')

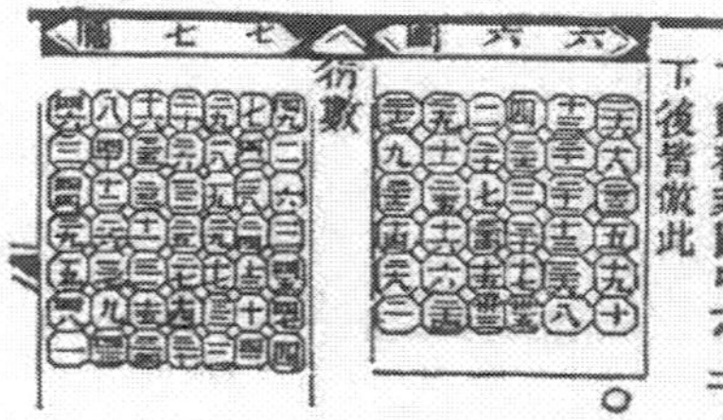

27	29	2	4	13	36
9	11	20	22	31	18
32	25	7	3	21	23
14	16	34	30	12	5
28	6	15	17	26	19
1	24	33	35	8	10

Above: A Chinese magic square of about A.D. 1590.

DHARANA : Skt. Concentration of consciousness by fixation of the mind & attention on one spot. Usually putting the whole mind on the tip of the nose, the naval or one particular idea or object. The purpose of Dharana practice is to still the mind as a preparation for Dyana or meditation.

DHARMA : Skt. The essential quality of the living being to render service or occupational duty.

DIANA : an ancient Italian deity, Goddess of the Moon & of hunting & Protectress of Women. Greek version is Artemis. p 156.

DIGITAL SUMS : e.g : the No. 108 has its single digits added = 1 + 0 + 8 = 9 . Sometimes called a Reduced Number, this study-thesis in the realm of the Multiplication Table (see Chap "X") has led to the Art of Atomic Structure . see p 102 & pp 173 - 176.

DOSSIER : bundle of documents on the same subjects.

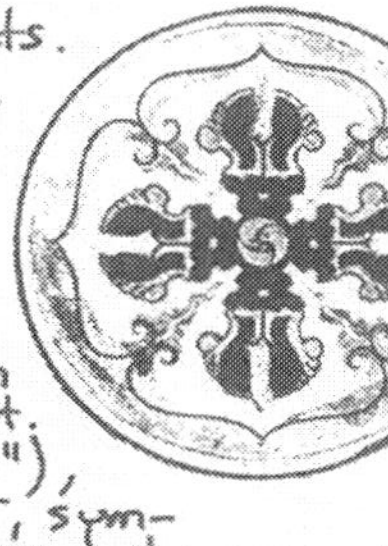

DOUBLE DORJE : Skt. Vajra, Tibetan-dorje. "Stone Lord" - King of Stones, thus "diamond sceptre", a symbol of sovereign power at least 3,500 years old. It is the double thunderbolt as held by Classical Jupiter, Hellenistic Zeus & early Indus Deity Indra. It symbolizes the total process of the Universe : dispersal, absorption, reintegration & , completing the cycle, the condensation in the central dot. From this central BINDU (Drop, Zero. - Tibetan "thig-le"), representing the seed of the Universe, the lotus grows out, symbolizing the polarities, the 4 directions : space, time, energy & form. The Double Dorje is a multi-dimensional mandala (meditation circle) going round in a circle & out around the viewer.

DOUBLE HELIX : Part of the intricate geometry of genetic coding existing as polynucleotide chains that are twisted together & held together by hydrogen bonds between the bases, which are anchored by a sugar-phosphate background. The bases adenine, guanine, cytosine & thymine bond the nucleotides in this spiral DNA molecule. (Keys of Enoch) .

DUALS : Two Platonic Solids are considered Dual if one shape exists within the midpoints of other shapes. pp 178 , 274.

DURGA YANTRA : Skt. a sacred diagram of triangles & lotus petals that actually represents the abode of that Shiva-related Goddess . p94 fig 71.

EAGLE : A symbol of height, of the Spirit of the Sun & of the spiritual principle in general. In the Egyptian hieroglyphic system, the letter "A" is represented by the figure of an eagle standing for the warmth of life, the Origin , the day .

EIGHTH CHAKRA : The template through which the higher energy systems encircling the body are unified with the human bio-computer circuits regulated by the energy flow of the 7 CHAKRAS. It is the Divine Light (Halo) seen around the Head uniting the human biological complex with the Overself Body.

ELEVEN : ELEVEN 11:11 a pre-encoded trigger placed within our cellular memory banks prior to our descent into matter which, when activated, signifies that our time of completion is near. p 132. New Age phrase coined by Solara.

ELOHIM : Heb. (sing. "Eloha"). The plural splendour of the Creator God as "pluralis excellentiae". The Creator Gods/ Divinities of the YHWH who control the calibrations of Light necessary to evolve all the combinations of the Image & Similitude through the Eternal Eye of the All-That-Is . (Keys of Enoch).

ENNEAGON : 9-sided polygon . see pp 65 & 66.

E PLURIBUS UNUM : Lat. Lit. One out of Many. Unity out of Multiplicity. see p 234. This is a good analogy of the LO-SHU !

ESSENES : a Spiritual Xtian brother/sister/hood into which Yesu, Isu, Iesus or Jesus was born into. Known as Nazarites, Nasirs or Yessenes, they practised community of goods, vegetarianism + fasting in their desert lands where they trapped dew & made exquisite gardens. Their bibles were full of m. sq. Parables of the Divine Kingdom. see p 125.

EURHYTHMY : a form of speech made visible by a sacred geometrical dance e.g : a large pentagram star is drawn upon the floor & all the angelically robed dancers follow this symmetry whilst the Poem is being recited. Developed by Rudolf Steiner (1861-1925) Austrian philosopher & founder of Anthroposophy : Anthropo = human being , sophy = Wisdom.

Peter Flötner — about 1546 — Mauresque with scrollwork.

FIBONACCI : circa C12th, his famous rabbit-growth experiment led to the universal natural series of Numbers : 1, 1, 2, 3, 5, 8, 13, 21, 34, 55, 89, 144 etc, increasing at the rate of approx. 1:618. Remembrance of this Fibonacci Series will unleash more Sacred Geometrical Codes. Commit it to your Memory Banks Now. It taps into the World of the atoms, stars & Nature. pp 111 + 246.

Fibonacci

FIFTH DIMENSION : The next "garment of light" that our matter-energy body enters, in the process of spiritual evolution. (K.O.E.).

FINAL CONNECTING LINE : of a M. Sq. Yantra, the last line drawn that connects the last No. to the first No. See "ALPHA + OMEGA".

FIXED DESIGN : "... means you have in your Life Purpose a fixed design. This design is what you set out on Life's Journey to achieve. To realize the growth of the sum of your total Experience through many Lifetimes. Yours is to express to others the Beauty of the Spirit, the Freedom of the Spirit & show that this Beauty can be for all men & women everywhere... Believe in your own power & the love that backs you. Have in your mind Growth of your SelfLove. Value the things that make you Unique such as your Beauty of Mind & Body, your own Freedom of Mind & Body, your own Humility". p 196. (This was a channelled message from a ClairAudient : Cameron to Jain circa 1978, age 21.).

FOCCI : plural for 'Focus', meeting point of rays, light, heat, or any other radiation. see p 143.

FREEDOM : "You may chain my hands & shackle my feet; you may even throw me into a dark prison but you shall not enslave my thinking because it is free". Kahlil Gibran.

F.R.E.E.D.O.M.S. : acronym for the living Sacred Geometry Mystery School that I have created. Its spiritual symbol is the m.sq. of 3×3 at 4 × 45°. Its aim is to teach children the connection of m. sqs to Atomic Structure & Free Universal Energy (see "SEARL, Prof"). The letters stand for : "For Research Expressing Essential Data Of Majik Squares". pp 8, 65, 66, 188, 231.

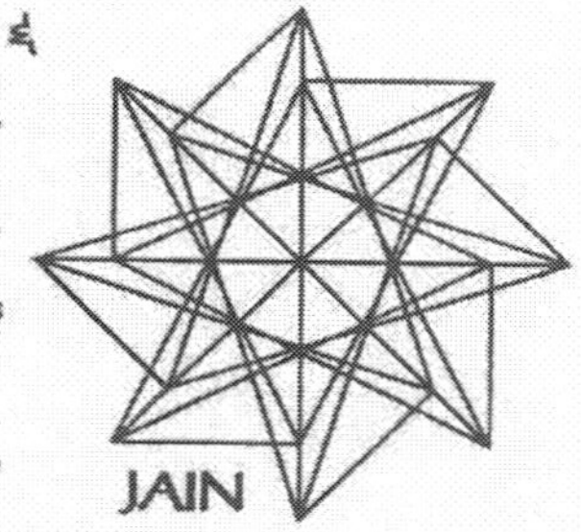

FREE ENERGY : clean, non-polluting energy to run our cars & power our homes based on Universal Spiral & m.sq. Harmonic Grids. see "SEARL, Prof".

FRET : an interlaced angular design of bands within a border. see p 56 fig 40.

Imitation of Saxon letters by Silvestre, 19th cn.

GANESHA : Skt. Elephant-headed pot-bellied God invoked as the Remover of Obstacles & Tantric master. His symbol the SWAZTIKA (卍, 卐) is buried under the first cornerstone of a newly built temple or house as a sign of blessing or auspicious beginnings. The m. Sq. of 6x6 has an essential Yantra that appears as the 卍!

GAYATRI : Skt. lit. "hymnal". Tat savitur varenyam bhargo devasya dhimahi dhiyo yo nah pracodayat = Let us contemplate that most excellent light of the divine Savitri (the Solar God) so that He/She may inspire our Visions.

GEMATRIA : The science of determining the necessary input power needed to build a consciousness body; i.e the mathematical calculation of the weights & measures supporting each consciousness vehicle. (Keys of Enoch).

GIGA : Gk. lit. Giant ("Gigās") = one thousand million = 1×10^9. Giga Consciousness.

GODDESSENCE : God.dess.essence. A Word coined by Jain to balance & merge the gender form/name of God as the essence of God (♂) + Goddess (♀). Keeps the Peace.

GURU : Skt. The syllable "gu" = darkness; the syllable "ru" = destroyer of that darkness. Light. He/she is the true guru by whose very contact there flows the supreme bliss (ANANDA). See "ANAHATA GURU".

GURU VIDYA : Skt. Sometimes called Vidya Gupta, referring to a secret transfer of traditional knowledge by word of mouth, passes from Guru to Guru via discipleship & these oral teachings are only Given when the disciple is ready.

Gorgon According to Frobenius, the gorgon is a symbol of the fusion of opposites: the lion and the eagle, the bird and the serpent, mobility and immobility (as in the swastika), beauty and horror. Hence it is symbolic of conditions beyond the endurance of the conscious mind, slaying him who contemplates it. Like other fabulous entities, it is also symbolic of the infinite number of forms in which creation can manifest itself.

GWALIOR : a place in India where a M. Sq. carved in stone is represented over the Gate or entrance. pp 42 & 43

HAHA-AHA : the representative incarnation of the Jain Sect disguised as Its own devotee, as Jain's (the author's) Higher Self & teacher of the magik Squares; who descended to teach Love of & Union with God through the process of congregational dancing, chanting, Theatre of the Holy Numbers, & discourses on the Maha YANTRA (Maha = Great, Yantra = Pattern) ⟶.

6	1	8
7	☆	3
2	9	4

Haha-Aha's intention is to assist Jain to spread M. Sq. Harmonics to all quarters of this World, on the surface & to the civilizations within; in all languages of this World, via books, video, CD Roms, TV, radio, satellite, telepathy etc. See "AH" & "AHA".

HAKIM : Ab. wise-man or sage; the traditional physician & philosopher as well as master of most of the other sciences.

HARA : The center of gravity of the body, situated between the SVĀDHISTHĀNA & MANIPŪRA CHAKRA. In T'ai Chi, it is the region about "1" below the navel, known as "the Cauldron".

HARMONY : Gk. "Harmos" = to join. That was one reason why magic Sqs. became known as 'Harmony Sqs.' (as in Vol 1), the concept of the book being to Join The Dots of the M. Sqs. in sequential order. The other reason for the name change being that some Christian schools reacted to the word "magic" as if it meant something 'diabolic'. Christian missionaries residing in Indian called these M. Sqs. "Diabolic Squares".

HEART : 'Hrid' in Hindu, the Spiritual Heart is celebrated by yogins & mystics as the seat of the transcendental Self. This secret cave is the place where the yogin must restrain his/her mind.

HECATE : Gk. Goddess of Moon, represented in triple form, having power on earth, sea & in the heavens. see p 156.

HERMES : or Thoth. Ibis-headed Egyptian builder of the Pyramids & creator of our ABC & 1,2,3. see p 138. Known as the MessenGer of the Gods. Carried souls of the dead to the underworld. He was the link between the extrahuman & the terrestrial. The 'perfected man' of the SUFIS. Invented the lyre. His flute cast a giant into sleep. His female double "Sesheta" built temples & preserved books of wisdom.

HETERODYNE : of Radio: a method of receiving continuous wave radiotelegraph signals by impressing upon the continuous radio frequency oscillations another set of radio frequency oscillations of a slightly different frequency. The heterodyne effect is the interference resulting in fluctuations or beats of audio frequency. p 110. 'The heterodyne children of light...'

HEXAGRAM : see "I-CHING".

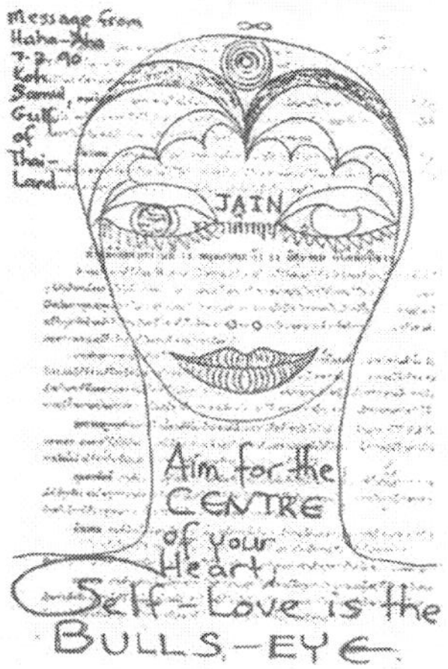

HIR : abbreviation & combination of "his and her" coined by Timothy Leary. see p 120.

HOLOGRAPHY : a form of photography in which no lens is used & in which a photographic plate records the interference pattern between two portions of a laser beam, to create a 3-Dimensional effect. see MAGIC CUBE.

HU or HOO : Ab. Sound of God. Herme's Word of Creation, resonated by Sufis. (the Order of the Whirling Dervishes).

HUM SAH : Skt. The Great Swan (The Self). MANTRA for Ajapa Pranayam meaning "I Am He (God)". Ajapa = repetition of the same sound. Pranayama is a breathing technique; as you breathe in mentally say: "Hum", as you breathe out, say: "Sah".

I AM THAT I AM : Heb. "Ehyeh Asher Ehyeh". The highest statement that a mortal can use in this world. It expresses the "covenant" between the human self & the Christed Overself, & a knowing of one's true identity, one's destiny & the keys to the higher thresholds. A holy MANTRA/salutation working with the Brother/Sister/Hood & Hierarchy of YHWH. (K.O.E).

IBIS : The ibis is related to Thoth, the Egyptian God of Wisdom. According to the Greek scholar Aelian, in his 'De Natura Animalium', the bird was chosen because it tucks its head under its wing when it sleeps, so that it comes to resemble the shape of the heart. Its stride measures exactly a cubit which was the measure used in the building of the Temples.

I CHING : Book of Changes, China's ancient oracle for 3,000 years. Based purely upon the M.Sq. of 3 → or LO-SHU. Its symbolism is based upon the 8 Nos. of the M.Sq. that surround the central cell of "5". These 8 Nos. are represented as 3-lined glyphs composed of either male, yang, full lines or female, yin broken lines. These correspond to Directions, Seasons, Father/Mother/Son/Daughter etc. When the 8 Trigrams are squared, 64 Hexagrams (6-Lined Glyphs) are permutationally created. The 64 binary codes equate exactly with maps of DNA structure.

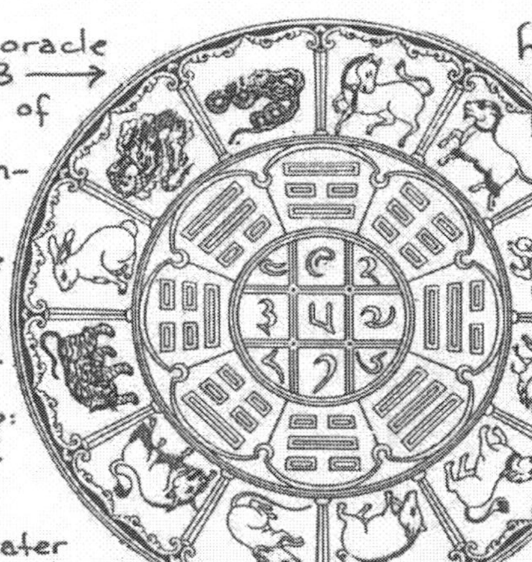

ICON : picture, image or other representation. e.g. of some sacred personage X^t, angels or saints. Iconagraphy. p 56.

ILM AL-'ADAD : Ab. Science of Numbers, later often used interchangeably with "ílm al-hisāb" = reckoning.

I.N.R.I. : Lat. Lit. "Igne Natura Renovatur Integra" which translates as "The whole of Nature is renewed by Fire".

ISHQ ALLAH : Ab. "God is Love".

ISOLATING : in regards to M.Sq. YANTRAS, it is the creative process of extracting an aesthetic center or section from the main diagram, that may be cut out, coloured-in, TESSELLATED or meditated upon. see p 18.

IS-RAEL : meaning the Crystal or pure white race, the first race that inhabited the world... a pure Light race... meaning ray or beam... From this race sprang the Aryan Race. see p 169.

JAINA : referring to the Indian Jain's M.Sqs. constructed 2,500 years ago at the time of MAHAVIRA. Jaina squares are shown on p 42-45.

JAINISM : Spiritual tradition founded by Vardhamana MAHAVIRA, an older contemporary of Gautama the Buddha, C5th BC. The Jain sect is very fanatic & ascetically rigorous. Their main virtue is AHIMSA (non-harming) & the practice of AMAROLI. One sect walks naked: "SKY Clad". The Jains have channelled through the M.Sqs. as intellectual pasttimes. (When I took on the legal name "Jain" in 1984 I had no knowledge of the Jains! This shows that all Knowledge is in our cells. see MAHAVIRA.

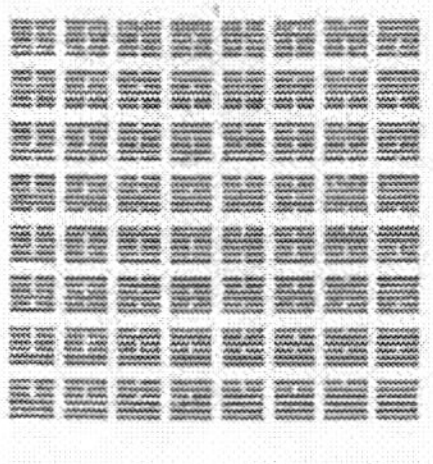

regular arrangements of the 64 hexagrams

JINN : Ab. lit. genius. A being from the Sphere of Mind, loving art, knowledge, beauty. Closer to man than the angels.

JIVA : the soul or atomic living entity.

JNANA : Skt. Wisdom, knowledge, intuitive insight, gnosis. A degree of absorption beginning with one-pointed concentration of an object to the exclusion of all other thoughts or awareness of externals & ending in a state of ultra-subtle residual perception. JNANA-MARGA = the path of Wisdom. JNANI = one who is engaged in the cultivation of knowledge.

JNANADEVA : boy-king-yogin-genius circa C13th died at 21, in a state of rapture. wrote great poetic works on spiritual philosophies. India.

JUNTRA : A M.Sq. printed or engraved on copper & worn over biceps as a sign of spiritual strength & harmony. This special M.Sq. often adds up to 20. ∴ called BISA YANTRA. p 93 fig 68.

KA : Egyptian word for Spirit, particularly a Soul being guided through the Underworld.

KAABA : a large black cube that millions of moslems worship by circumambulating it. It is the tomb of Mohammed their Saviour. The worship is geometrically "The Mystical SQUARING OF THE CIRCLE". see pp 122, 123. see KORAN.

KALI YUGA : Age of Darkness & Ignorance that we as a planetary community are in now. Kali is one of the Goddesses of Destruction, synonymous with Durga. Last age in the cycle of a Maha-Yuga: (Great Age).

KARMA : The Law of Cause & Effect. The moral force of one's intentions, thoughts & behavior. Connects with the idea of rebirth. By constantly cultivating a self-transcending disposition, the viscious cycle of karmic existence can be intercepted.

KETHER : Heb. lit. Crown. The topmost Sphere or Globe of Consciousness on the QABALAHL t/c Tree of Life designating Heaven or AIN SOPH AUR. The Primordial Point, first impulse toward manifestation. p 114.

KITAB AL MALIKI : Ab. lit. The Royal Book.

KNIGHT'S MOVE : L-shaped moves, as in chess, e.g: corresponding to 2 cells up & 1 cell to left. p73 fig 52.

KORAN : Sacred scripture of Islam believed by orthodox muslims to contain revelations made in Arabic directly to Mohammed. The Muslims mastered Maths, M.Sqs. & Abstracts as it was decreed by the statement : "Come, let us make Man in our Image". That human form was holy & ∴ Life-drawing was forbidden. see KAABA.

KOSMOS : Gk. lit. Order. A study of the Cosmos involved a study of Number especially in PHI ratios. p 234.

KUNDALINI : Skt. Serpent power residing at MŪLĀDHĀRA CHAKRA at base of Spine, brilliant as a million suns. In its coiled state, it is potential energy awaiting for one day in your life to ascend along the central conduit of the body. When this psycho-spiritual energy reaches the locus of the Crown (or 1000 petalled Sun) it floods the entire body & body consciousness with divine nectar AMRITA. The body is thus experienced as the Body Divine. The wise control this force while those who realize it ignorantly are destroyed by it.

LALITA : Mother Goddess that is worshiped in India. Resides in the center of the holy SHRI YANTRA, BINDU. p 96.

LATTICE OF DOTS : e.g: a 3×3 Square of dots. Each dot symbolizes a No. of the m. Sq. required to transform No. into Art.

LAY-OO-ESH : Heb. "Pillar of Light". Energy projection of Light used by the Brother/Sister/Hoods to communicate with the Faithful by using the Harmonics of Light. (Keys of Enoch).

LILA : Skt. Play, Pastime, the Divine Love Play of consciousness. see CAITANYA.

LOOM : see p 188

LO-SHU : According to legend, the Lo-Shu was a "heavenly Gift" to the Emperor Yü the Great. The Lo-Shu or m.Sq. of 3×3 was brought by a turtle which emerged from the River Lo C5th BC. The Cosmic Nos. appeared on the turtle's VENTRAL. The Chinese regarded Earth as Square & Heaven as round. ∴ the Circle of 8 TRIGRAMS surround the Square of the Lo-Shu that represents the 9 Provinces of the known world. These sacred Nos. are a key to understanding atomic / crystal structures + FREE ENERGY systems. see MING TANG.

Based on a Tibetan colour-print

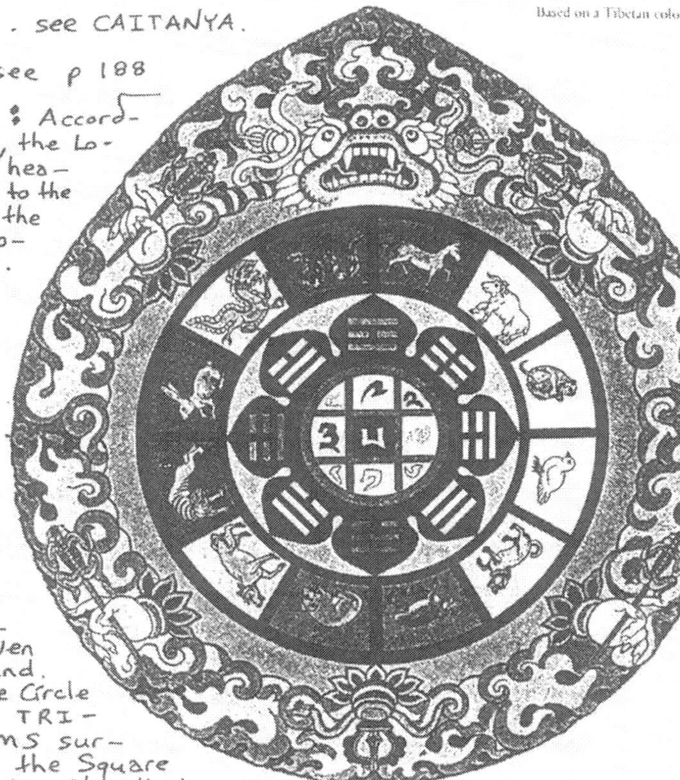

SINO-TIBETAN TORTOISE
WITH ZODIAC ANIMALS, TRIGRAMS AND MAGIC SQUARE

LINGAM : Skt. Phallus. Shiva's penis. A fertility symbol, creative force. Such stone carvings are blessed by pouring milk on the Shivalingam. The combined diagram of the YONI(∇) + Shivalingam is shown in this glossary under CHAKRA, under MŪLĀDHĀRA. These ancient cults worship the Essence not the organ.

LOZENGE : lit. "diamond-shaped", originally a small confection of sugar often medicated. On p 82 there appears a brilliantly executed diamond array of odd Nos. in the centre of the M.Sq. A masterpiece of design.

MADZUB : A God-intoxicated one.

MAGIC : the occult art of influencing one's environment through practices & rites based on Natural & Universal Laws. The word for spiritual perfection & paranormal power is the same in Sanskrit : "siddhi". Numerous siddha yogins have fallen from grace due to misuse of acquired powers for selfish ends.

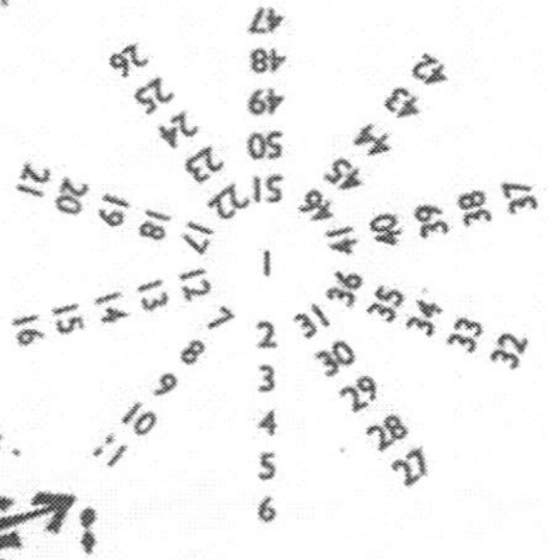

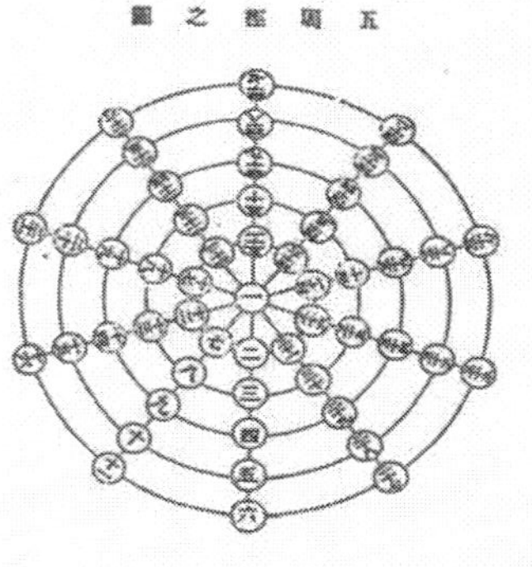

MAGIC CIRCLE→: a circular, concentric array of digits whose diameters are equal in their sums. Shown here is a Japanese magic circle from the works of Seki Kowa, C17th. Not only do the 5 diameters all equal 265, the 5 concentric rings also have a CONSTANT of 265. This rare & fascinating Magic Circle of 51 Nos. may also be called a magic Decagon (10 sides).

MAGIC CUBE→: All sums in 2D & 3D add up to 42. This magic sum or Magic Cube CONSTANT has a formula : $C = \frac{n}{2}(1+n^3)$. The 3-D YANTRAS produced by weaving a thread from ALPHA to OMEGA equates with realistic models of atomic structure. Info, not yet released.

A MAGIC SQUARE OF THREE—A CUBE OF THREE

OBSERVE THAT A MAGIC LINE CANNOT BE FORMED OF LESS THAN 4 NUMBERS, OR A MAGIC SQUARE OF LESS THAN 9 [FIG 1].

FIG 1.

8	1	6
3	5	7
4	9	2

A MAGIC CUBE REQUIRES AT LEAST 27 NUMBERS [FIG 2], AND EVEN WITH THIS NUMBER THERE ARE MANY LIMITATIONS DUE TO LACK OF SCOPE FOR NUMEROUS COMBINATIONS POSSIBLE WITH CUBES OF 5, OR BETTER, OF 7.

MAGIC CUBE OF THREE

EACH ROW AND EACH COLUMN OF EACH OF THE 9 MAGIC SQUARES (3 SQUARES TO EACH DIMENSION) ADDS 42.

EACH OF THE 4 CUBE DIAGONALS ADDS 42.

EACH OF THE 6 DIAGONALS OF THE 3 MAGIC SQUARES CONTAINING THE CENTRAL NUMBER 14, ADDS 42.

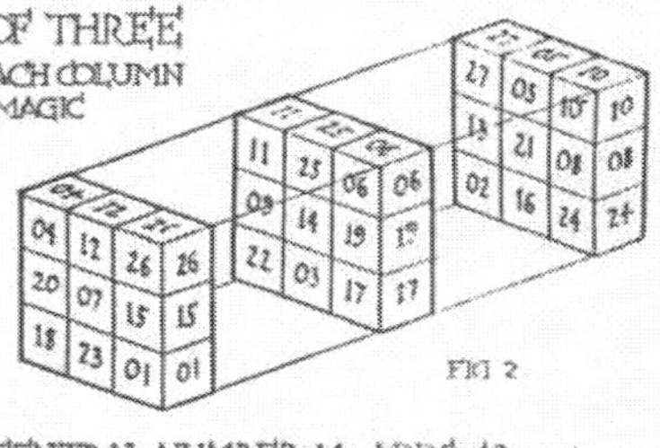

FIG 2

ANY RANDOM LINE DRAWN THROUGH THE CENTER OF THE CUBE WILL CUT TWO SURFACE CELLS WHOSE NUMBERS WILL ADD TWICE THE CENTRAL NUMBER 14. ALL THESE LINES ARE "MAGIC" LINES.*

MAGNITUDE : or Order of a M.Sq. is its size. e.g: A M.Sq. of 3 is of the Order or has a magnitude of 3.

MAHARISHI : Skt. "Great Seer". A title of respect for an adept.

MAHĀTMĀ : a great Soul, or devotee. see ATMA.

MAHAVIRA : Contemporary to Gautama the Buddha. He was the last of the prophesized 24 great JAIN leaders/teachers. Indian posters depict him in meditation with several M.Sq. number boxes surrounding him. p45 see JAINISM.

MALKUTH : Heb. The 10th Sphere or Globe of the Hebraic QABALLAH, designating the Earth plane of consciousness. Each sphere is attributed to a specific MAGNITUDE of M.Sqs. e.g: The M.Sq. of 10 relates to the plan/plane/planet Earth.

MANAS-CHAKRA : Skt. lit. "Mind Wheel". 6 petalled white lotus, esoteric center in the head situated above the 3rd Eye & called "Surya-Mandala" (Solar Orb) & a.k.a. Jnana-netra = (Eye of Wisdom).

MANDALA : Skt. lit. "Circle Orb". It is the love power transmitted into a design. In a state of concentration, the center of the devotee's own being is the Mandala. It is a psycho-cosmo mechanism for effectively relating the inner space. In the center is the BINDU our galactic heart. The surrounding rings represent various levels of existence. It is always a consecrated space. Some view it as the body of one's chosen deity, to become One with that deity via concentration. Many people think that a YANTRA is a mandala. The YANTRA is the 2-D sacred art or visible diagram meditated upon to understand the 3-D invisible holographic version e.g: Circle = Sphere, Square = Circle, & Triangle = TETRAHEDRON. See MOUNT MERU. The Mandala is a psychic complex which conditions the return of the psyche to its potent core. Hence the initiation process is often referred to as a "march towards the Center" so that the adept can interiorize the mandala in its totality, counterbalance the opposing dimensions projected in its symbolism & finally be reabsorbed in the cosmic space represented symbolically in the inner circle. The mandala aids in dominion over mental Processes.

THE ESOTERIC TRADITIONS HAVE HANDED DOWN TO US THE KEY TO TUNING IN TO THIS PLANE. IT IS THE MANTRAM OF THE HINDUS, THE NEMBUTSU OF THE BUDDHISTS, THE REPETITIONS OF KYRIE ELEISON OF THE HESYCHASTS, ORTHODOX HERMIT MONKS, THE WAZIFA OF THE SUFIS. SUCH TECHNIQUES DERIVE FROM AN EXPERIENTIAL KNOW-HOW THAT TEACHES ONE HOW TO ATTUNE THE FREQUENCY OF HUMAN CONSCIOUSNESS IN SUCH A MANNER AS TO ENABLE ONE TO BE RECEPTIVE TO VIBRATIONS OF A DIFFERENT ORDER FROM THOSE GENERALLY PERCEIVED BY THE EARS.

THERE ARE MANTRAMS IN SANSKRIT, IN PALI, IN TIBETAN, IN

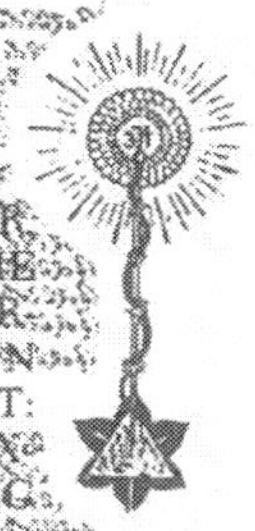

FOR EXAMPLE, IN AUM YOU HAVE THE A IN THE HEART CENTER, THE U IN THE THROAT CENTER, AND THE M AT THE TOP OF THE HEAD, WHICH IS VERY SIMILAR TO ALLAH HU, WITH THE DIFFERENCE THAT THE M IS DIRECTED UPWARDS THROUGH THE CROWN WHEREAS THE HU IS DIRECTED OUTWARDS THROUGH THE THROAT: AND IN HINDUISM THE WHOLE TREND IS TOWARD UNITY, THE EXPERIENCE OF ONENESS, WHEREAS IN SUFISM IT IS TOWARDS BECOMING A CHANNEL THROUGH WHICH GOD MAY MANIFEST HIMSELF IN MAN.

Rajayoga—Concentration of Thought

MANTRA: Skt. derived from the verbal root "man" = to think or mind & the suffix "tra" suggesting instrumentality or protection. ∴ It is thought or intention expressed as sound, or sound waves that protect the mind from wrong thinking. These syllabic hymns or prayers are associated with a particular invisible power or deity e.g: the Bija (Seed) mantra of Kali is "Krim", of Shiva it is "Hrim" etc. These sacred syllables charge the physical body with the powers & rapture of the Divine Mind, & give attunement to the Soul. Mantras represent Cosmic form at its most subtle, as vibration. The word is pronounced as "Muntra", & similarly Tantra is Tuntra, Yantra is Yuntra, like JUNTRA.

MANIPURA: The spiral center of psychic electricity which revolves around the solar plexus, having 10 divisions or petals. Governs the Gut feelings about reality & the logical sense of linear order. see CHAKRA pp 120, 157. Some Pundits refer to it as "The Seat of our Emotions".

MANJUSHRI: Main deity of the Tibetan calendar, positioned between Avalokitesvara, (Goddess of Love / Compassion) & Vajrapani who represents Power, in a Monster form. Manjushri holds a sword of Truth & scrolls of Knowledge & wisdom. He incarnates as the Fiery Tortoise revealing the sacred M.Sq. on his VENTRAL or underbelly, as the Centre of that Cosmology. see p6 + LO-SHU.

MAR CHARBEL: Ab. Mar = Saint. Lebanese monk who never touched money in his life. Spiritual healer about 70 y ago. Now accepted by the Roman Catholic Church as an authentic Saint. Mar Charbel is a spiritual Guide in my Life who revealed himself to me in Lebanon, 1979.

MATRIX: Lat. lit. Womb, Source. A 2-D rectangular array of Nos. That which gives origin or form to a thing which serves to enclose it. pp 11, 15.

MAYA: Skt lit "Ma" = not, "Ya" = this. The appearance of things, the magical illusory play of phenomena. Forgetfulness of relationship with The-All-That-Is.

MEDITATION: Skt. "Medha" means 'direct perception of wisdom'. The practice of systematically vacating & unifying consciousness to attain an unusual degree of lucidity. Staring at Yantras helps the mind from wandering.

MELCHIZEDEK: Eternal Lord of Light in charge of organising the levels of the heavenly worlds of YHWH for transit into new creation. ('Keys of Enoch' by J.J. Hurtak).

MERKABAH: A Spiritual / Divine / Light Vehicle used by the Masters & Mistresses to probe & reach the Faithful in the many dimensions of the Divine Mind. (K.O.E). The Merkabah can take on many forms of a brilliant BRIOLETTE in the physical worlds. (Ezekial 1:4-28 & 2 Kings 2:11). It is the ultimate & simplest Ascension Vehicle. Its primal form is the 3-Dimensional Star of Solomon known as the STAR TETRAHEDRON. see pp 152, 153. Refer also to "QABALLAH". The tetrahedrons are spun towards the speed of LIGHT.

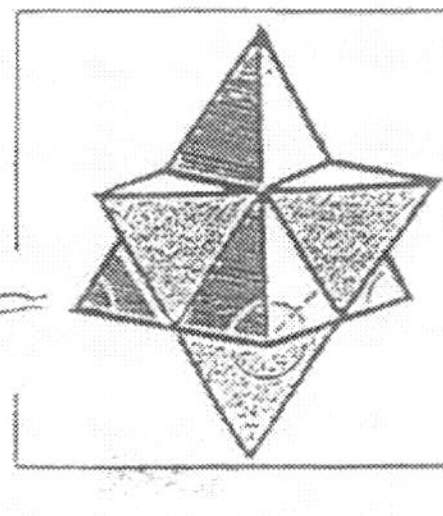

METATRON: Creator of the outer worlds & Creator of the electron. The resonant timeless geometry to assist in tuning into the consciousness of Metatron is known uni- & multiversally as Metatron's Cube. see p 249, 152, 153.

MING TANG: ancient Chinese agricultural rotational methods based upon the LO-SHU or M.Sq. 3 in Han Dynasty, 200 BC. The division of the Chinese Empire in 9 compartments specific to the Lo-Shu. Also found in early Islamic ALCHEMY. see p 6.

WORLD TRIAD

MITSU-TOMOE: or "Maga Tama" or World Soul or Triad, the Japanese triple form of the source of life (YING YANG) representing the elements of Fire, Air & Water. It is worn to protect the household & person from Fire, Flood & Theft. Its Talismanic origin is the 3-limbed SWAZTIKA Cross symbolizing ceaseless change & the eternally spiralling cycles of time. Image of Eternity.

MOD: abbrev. for "Module" or "Modulus" = Latin for small measure. A standard unit for measuring e.g: the 64 Nos. of the M.Sq. on p20 can be rewritten only in 8 units of "1, 2, 3, 4, 5, 6, 7, 0". Thus every multiple of '8' gets replaced or written as '0'. see p 20, 21.

MOKSHA: Skt. Liberation from human bondage & material worldliness by freeing the consciousness from its own creations. A shift in consciousness where one transcends all duality. Some yogins believe there is no moksha because we are already enlightened, all liberated!

MONTESSORI: Maria, né 1870-1952. Italian educational reformer. A spiritually-based school, like STEINER schools, where children learn self-education by themselves & practice topics like YOGA & MEDITATION. p3.

MOUNT MERU: The "Jewel-studded mountain-like peak" of the 3-Dimensional Mandala, abode of the gods. The Center of a 2-D YANTRA is really, in 3-D, a 1,000 petalled crown over the city, palace or island where the Self resides. see pp 95, 98.

MULADHARA: Skt. Root or perineum centre. see CHAKRA.

MUNI: Skt. Sage, etymologically related to "mauna", which means "silence". May every step you take be a dance & every word a song. Speak no triviality.

MYOPIA: short-sightedness, due to parallel rays focussing in front of the retina so that only near objects are seen clearly. see p 50.

NABI: Ab. a bearer of a particular message for a particular time. Prophet

NADA: NADAM. Ab. Inner Sound current of consciousness vibrating in the liquid crystal of life that is called a human being. Shown here is Milarepa listening to the Sounds of the Universe

MILAREPA

(Tibetan—Mi-la-ras-pa). The cosmic cotton clad one, the great Saint and Yogi who committed his life to the gaining of enlightenment. He lived as a hermit for many years in the Himalayas eating nettles till he exclaimed his hair was turning green. His Guru Marpa initiated him into the teachings of the Demchog Tantra—the Yoga of the Inner Fire (tum-mo-like the Bardo-Thodol). Marpa ordered him to build four houses with his own hands, the first circular, the second a crescent, the third a triangle and the fourth a square. In order to expiate his bad Karma he had to demolish each as he built it. This exercise made him aware of the psychic correspondences: circle—water—peaceful action; crescent—air—far reaching action; triangle—fire—powerful action; and square—earth—stern action.

He once said "Accustomed long to meditating on the Silent Truths I have forgotten all that's said in printed books." He is portrayed listening to the sounds of the Universe.

NASIK: Village in India where M.Sqs. were common & worshipped. A missionary living there, A.H. Frost, thus termed them: "Nasik Sqs".

NATURAL SQUARE: sequentially arranged box of Nos. in their natural counting order e.g: p27 fig 16. Prof. SEARL calls them "Uniform Squares".

NAVA-YONI-CHAKRA: Skt. YANTRA of 9 (nava) Vertices of interlacing triangles symbolizing the 9 cosmic wombs (YONI) p 95. Symbol of fertility & the Source.

NIRVANA: the extinction of the flame of "I" The end of the process of materialistic Life.

NIRVANA – THE UNCONDITIONED STATE, AT-ONE-MENT WITH EVERYTHING AND NOTHING, BEYOND KARMA. THE FOUR STAGES IN ATTAINMENT OF NIRVANA:

1. SOTAPANNA – LIT., STREAM-ENTERER.
2. SAKADGAMI – LIT., ONCE-RETURNER.
3. ANAGAMI – LIT., NON-RETURNER.
4. ARAHANT – LIT., SAINT, THE POINT AT WHICH THE LAST VESTIGE OF SELFISH MOTIVATION HAS DISAPPEARED.

Page 268

NOUGHT DEGREE: A non-superimposed or non-rotated m.Sq. pattern (0° YANTRA) drawn as it is. Drawn only once. The 2nd step into transforming mathematics into Art. The first step is to create the LATTICE OF DOTS which appropriates each No. with a specific dot.

NUMBER: a a concept of quantity that is derived from a single unit. Every No. occupies a unique position in a sequence, determining a hierarchy. b) Slang – a sexually attractive girl. c) "to have someone's number" is to discover a person's true character or intentions. d) "one's number is up" i.e dead, ruined. e) "Your days are numbered" f) "Number One" = a phrase for oneself, myself. g) "Numbers" = 4th book of the Old Testament, recording the numbers of the Israelites who followed Moses out of Egypt h) "number" comes from the Old French: "Nombre" nb 'nom' = name – "What is your Name?", "What is your Number?"

NUMINOUS: a feeling of awe & wonderment characteristic of Man/Woman's communion with GODDESSENCE. p 221

OM:

OM or AUM:

There are 4 measures or parts composing OM: Pronounced as AUM & the dot which signifies a nasal humming

om. The sacred monosyllable *om* is the oldest and most venerated of all *Hindu **mantras*, and is also employed in *Buddhism. It is also called the "root"—or *mūla-mantra* and often precedes other *mantras*. The *om*-sound symbolizes the *Divine. The **Maitrāyaniya-Upanishad* (VI.22) refers to it as the sound of the soundless *Absolute. It is by means of this numinous sound that the **yogin* focuses his *attention to the point where he can transcend the finite *consciousness in its entirety. The above *Upanishad (VI.24) likens the *body to a bow, the syllable *om* to an arrow, the focused *mind to the arrow's tip, and the ultimate Mystery as the target.

48. The sacred syllable om.

Universal Sound or endless and beginningless Pure Mantric sound of empty space which merges and emerges from the still silence of Pure Consciousness. The vibration at the back of the cosmos transcending any definition and beyond the boundaries of thought. It is activated by the motion of energy through space just as breath activates the sound in the column of air in the flute. OM is that which manifests any thought which arises in consciousness as vibration; e.g. as in OMNI-science or OMNI-presence. OM is everywhere vibrating supersonic overtones throughout the whole of creation. This sound symbolically represents the perfect wholeness of things. The sacred syllable usually written in modern Sanskrit as ॐ or in ancient Sanskrit as ओ

OM TAT SAT: transcendental syllables chanted by BRAHMINS during sacrifice.

ON EST: Lat. lit. Being one with that which is. Honesty.

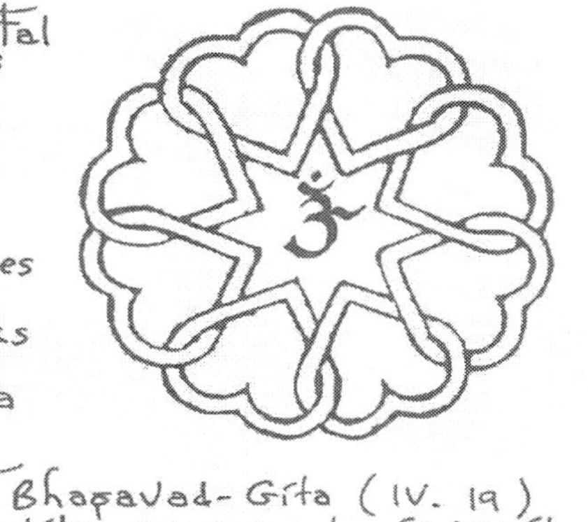

PAN: Gk. God that rules over the countryside, pastures, forests, flocks of sheep. Half man + legs ears & horns of a goat. p 156.

PUNDIT: Skt. Scholar. The Bhagavad-Gita (IV. 19) describes the true pundit or pundita as one who is devoid of desire (kama) & motive.

PARTICIPATION: is sharing, thus caring. One way of healing the rampant Split Participation in our Society is to sincerely learn this: that When You meet Anyone, Remember, It is a Holy Encounter.

PARABRAHM: Skt. p222. The Ultimate God.dess. essence Presence that dwells within.

Paradigm = a pattern, an example. p 156. e.g: magic Square Paradigm.

PARAMAHANSA: Skt. lit. "the topmost swan" The topmost class of devotees. The swan-like etheric energy shape that forms around an enlightened person's crown due to the release of an endomorphin that is sublimely blissful & AMRITA – like.

PHI: Gk letter of alphabet, symbol "Φ" represents fertility, growth, coitus. "O" = womb, "I" = penis. Phi is the most elegant & aesthetic proportion used in Nature, human anatomy, temple building, famous artists & even in magik Squares! Unity is divided by ·618 repeater (i.e it is an irrational No. that has no end). Can be expressed as 1 : ·618 or 1 : 1·618. Also, 'Golden Section', 'Sacred Cut', 'Divine Proportion'.

PHILOSOPHY : Gk. "Philos" = love , "Sophia" = Wisdom, ∴ = Love of Wisdom.

PINGALA : A psychic conduit which channels the right-hand energies of Shiva force around the central Sushumna. Prana or life flows through the PINGALA NADI and terminates in the right nasal sinus approximately every 55 minutes. It is the positive and opposite polarity to the IDA channel.

PLEIADES : Star cluster of 7 visible suns known as The Seven Sisters, or The Jewel in the Night Sky. There are 22 visible under telescope. It is our galactic center, in that our solar system takes approx 26,400 yrs to cycle / spiral around one of the stars named Alcyone. p 103. The modern esoteric 'knowing' or belief is that our solar system is the 8th Star of the Pleiadian cluster, ie: we are actually part of the Pleiades!

POLYGON : Gk. = many angled. A closed plane figure having more than 4 sides or angles as shown here →

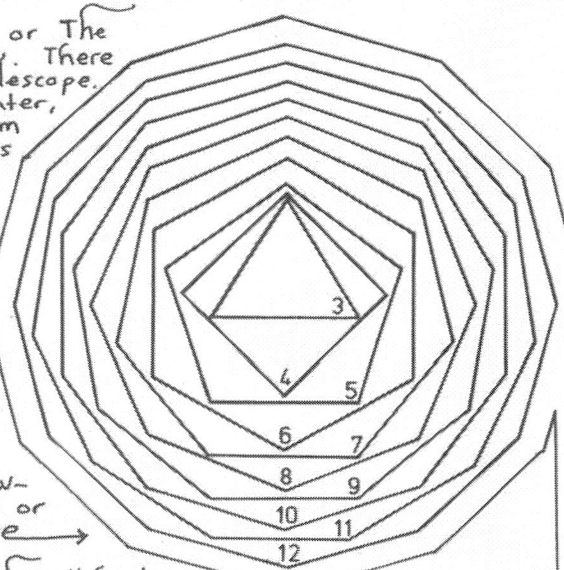

PRAKRTI : Skt. the ultimate female principle. p 97.

PRANA : Skt. The vital energy of the cosmos, carried by breath. The universal life-force, the vibratory power that underlies all manifestation. This Prana is meant to circulate in or along 72,000 channels (Nadi) that feed all the organs of the body. Prana establishes links between breath, consciousness & gateways to the heavenly world.

PREMA : Skt. the highest stage of love of God.

PSYCHOMETRY : "Psyche" = Soul & "Metron" = to measure. ∴ it is the measuring of the soul of things. The reading of objects, or rather their natural vibratory energy, by handling them.

PUJA : Skt. The daily worship of one' chosen deity by the offering of fruits, flowers, honey, jewels, scents etc or washing of feet. Sometimes MANTRA is chanted. In Rajasthan, Pushkar town, many buisnesses offer puja to magic squares framed in alcoves or recesses of walls, for the invocation of successful buisness. see p6

PURUSHA : Skt. the ultimate male principle. The feminine polar opposite is : PRAKRITI, meaning Nature or Creatrix.

Supreme Consciousness of Self without attachment to "I" or Self Consciousness. The first thought or desire of anything before it manifests in our consciousness. That which sets going the creative force and creates forms although it is itself Pure and formless. Like electricity in its three forms of protonic, electronic and neutronic energy but as yet unmanifest as in the formlessness of light and radiation.

PYRAMID : In Phrygian / Gk. "Pyr" = Fire , "Ur" = light ∴ lit. "The Fire in the middle". In Hebrew = "Urim-middin" = light measures, in Semitic "urim" = the lights. Light is formed of pyramid-like sub-atomic particles that have an angular velocity of 144,000 which is the light value in grid seconds. The triangular faces of the pyramid represent our Emotional, mental & Astral bodies. When these are aligned, or in balance & awakened, a 4th Dimensional experience occurs as symbolized by the 4 sides of the base. The Keys of Enoch (K.O.E) defines "pyramids" as time clocks or time-vaults built in various sectors by earlier rounds of higher intelligence for watching or directing the destiny of an experiment. Earth is a vast library of DNA data-bank. Coded time-structures. I see the Pyramids as celestial monuments for the Living, not as tombs for the dead.

QABALLAH : Heb. The science of the many universes of higher intelligence, hierachies & theophanies that serve the Godhead. Explains the physical creations of the worlds with colors, planets, magical squares & sacred names placed onto a Tree of Life. A trinity of triads with 10 Spheres or SEPHIROTH. Divine Prayer written in the shape of human form, which is the reflection of God-Man : Adam Kadmon the primordial man. see P52

YA RAHIM : Ab. All encompassing mercy & compassion. Com- Passion. Come Passion.

YA RAHMAN : Ab. Divine creative Love.

RANGOLI : sacred YANTRIC floor patterns at entrances of Indian houses consisting of : cont.

wet rice flour or other grains. Hindu women create these designs on auspicious days see p 94.

RASUL: Ab. Messenger, bearer of a universal message.

RECRUDESCENCE: breaking out afresh into renewed activity, revival, reappeara in active existence. p 156

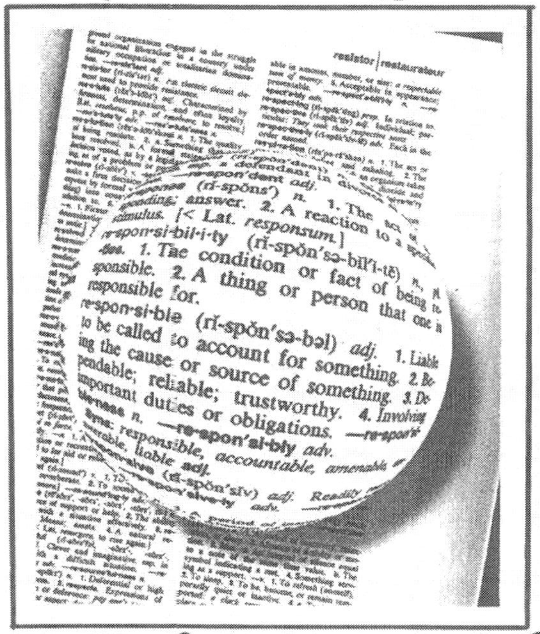

RESPONSIBILITY: is the ability to be able to respond.

REVOLUTION: = re – evolution.

RISHI: Skt. seer title of the Vedic bard who "sees" the hymn before composing it. A female saint is bestowed the title of "rishika". see p 234.

RIVER of CRYSTAL: The "milky Way" galaxy into which the Lords of Light "dip" their thought-forms. A stream of myriad star creations which has its own "free path" with "edges" that can be conceived of as "banks" for planting a variety of life forms. (K.O.E)

RODLIN COIL DESIGN: winding pattern for a torus which creates synchronized electricity. See diagram above on Right. The "magic" of these geometric windings that create a doubling of output, are similar to the Magik Square YANTRAS. By winding coils in this manner, you orient the electrons moving through the windings & thereby minimize random collisions of electrons, as well as heat friction & reluctance. Notice how the cross-over of straight-lined electrical geometries form a distinctive "MAGIC CIRCLE". Imagine electrical circuit boards imitating these majic square Yantras. Also read "ALPHA & OMEGA".

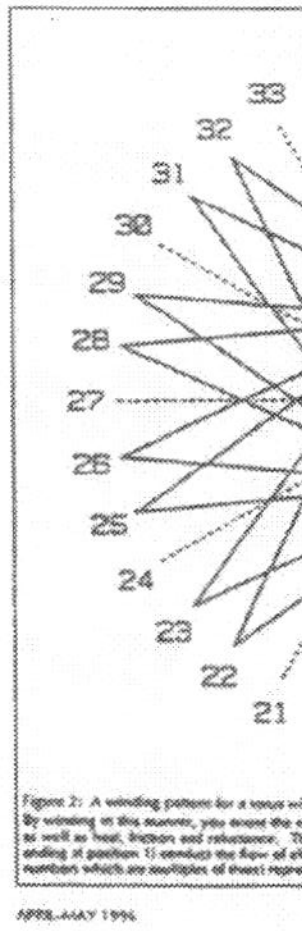

RUH or RUACH: Spirit, the Divine Center of the Cosmos, that level of existence between the Origin & the Universe corresponding to the "Spiritus" of the ancients. In medicine, the spirit descends upon the mixture of humours & is the subtle body between the physical body & the force of life from the world above. Nb: The 'Spirit' not the 'soul'. The Soul is the natural consciousness of life without the higher realization of worlds or rebirths that have preceded this world of worlds that are to come. (K.O.E.) Ultimately, the soul intertwines with the Spirit or OverSelf or Ruach.

SABI: aesthetics within poverty. A Japanese word that suggests correct use of humble materials is more meritorious than opulence.

SACRALIZE: to make Sacred p 140.

SADGURU: Skt. A teacher who reveals the true Self.

SADHAKA: Skt. A competent ardent practioner of the spiritual practices and disciplines (Sadhana), continues with YOGA, is well-informed about the shastras (teaching), is well-spoken, healthy, avoids public gatherings. He-She may also be known as a "Sadhu" or saintly, virtuous person.

SAHASRARA: or **SAHASRĀRA PĀDMĀ** or Fontanelle see CHAKRA + p 157. Crown Center - 7th Chakra.

The highest Vibrational spiral of psychic electricity at the top of the spinal column above the Pineal Gland. It is often referred to as "The Thousand-petalled Lotus". On babies under 3 months and Saints this spiralling field of bio-psychic rays of subtle pure consciousness appears as a golden yellow nimbus or halo of psychic electricity. Once Ego enters, the nimbus shuts off and the Sahasrara becomes the seat of Self-consciousness. The Selfless consciousness arouses the Shiva force in union with Shakti force and the whole body is filled with light of consciousness.

SALAAM: Ab. Peace

SAMADHI: Skt. The state of trance when the mind IS Pto.

GLOSSARY

THE COSMIC GLOSSARY CONTINUED

SAMADHI: skt, the state of trance when the mind is fixed on the Supreme. Enstasy arrived at by the perfect forgetting of the state of meditation which precedes it. Subject & object into a state of oneness with all. In the Hatha-Yoga Pradipika (IV. 188 ff): "The yogin yoked by enstasy does not experience cold or heat, sorrow or joy, honour or contempt ... & is immune to magical influences from mantras & yantras & is invulnerable to any weapon & unassailable by any person". It is the condition where the yogin realizes: "I AM the Absolute".

The state of consciousness which identifies with the Supreme Superconscious without any separation of "I AM". A union of "I" in the individual with the universal so that knowledge is no longer personal. The annihilation of the human personality—its identification with an experiencer of Self. All selves become ONE SELF in the highest state of existence.

SAMSARA: skt, the vicious, repetitious cycle of birth & death & rebirth.

SAMTA: skt = Equality on all levels.

SANYASIN: skt = a renunciate p 43

Yoga meditation: the esoteric fire-rite (yajna) of union with spirit

SARASWATI: Hindu Goddess of music often depicted holding a sitar. see p 36. Also, consort of BRAHMA the Creator God. As his SHAKTI, she is endless fertility & is the power of materialization. Saraswati's vehicle, the peacock, symbolizes in the spread of its variegated plumage the miraculous manifestation of objective creation from the Cosmic Void. see p36.

SAT: skt. Eternal Existence without form or location in space.

SATORI: skt. Sudden enlightenment & all-Knowingness.

SATSANGA: skt. "Sat" = Truth, "Sanga" = union, congregation. ∴ it is the gathering of people seeking the Truth; reading Scriptures.

SATYA: skt = Truthfulness. This is held in high

regard in the spiritual traditions, for without it, all practices are futile.

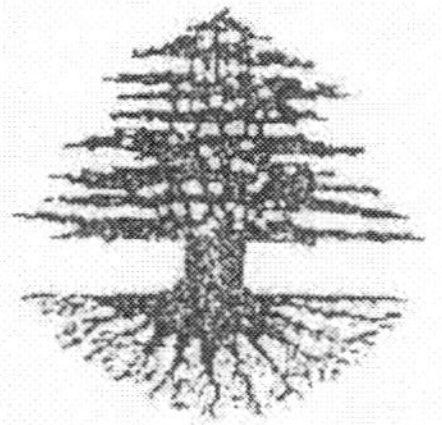

SEARL, PROFESSOR JOHN. R: A creator of Free Energy generators that can power cars & homes for free, with clean & non-polluting systems. He was the creator, in 1971, of levity discs & anti-gravity devices. Made of magnets, rollers & metal powders that correspond to the Periodic Table of the Elements, his work on UFO technology actually relates directly to the mathematics of how magic Squares are constructed. Prof. Searl's genius thus validates my life's work, in that m. Sqs. do have a practical use. m. Sqs. = Free Universal Energy. His many books are called: "Law of the Squares". Please write to me for more info + contacts on this subject. That is why m. Sqs. must get out to children, as even kindergarten children can vibrational absorb the contents of this lost knowledge. As a mystic & scholar, I know that m. Sqs. relate to Atomic Structure & Universal Principles of Eternal Cosmic Energy, for I have studied this for 20 years. Subsequently, when some knowledge is brought forth to Earth from Dimensional Realms Above/Within, the "inventions or ideas" come often to 2, 3 or 4 or more people simultaneously & in different countries, as it surely did with John Searl & myself. News Release: The "authorities" have destroyed, by fire, Searl's books & home & he lives under another name, so the rumour goes. So Now is the time to act. Buy our books & teach it to your children. End this conspiracy on the true Sciences.

SEMI-MAGIC SQUARE: when all else adds up to the same CONSTANT except for the Diagonals e.g: fig 127 on p 146.

WHEN ENOUGH PEOPLE LEAD, THE LEADERS WILL FOLLOW

SEPHIROTH: Heb. the ten qualities of the mystical Tree, set forth as a major doctrine of the QABALLAHistic "Book of Splendour" (Sefer ha-Zohar) published in 1280 AD. The Mystical Tree stood for the "World of Union", showing the process of life flowing from divinity into the whole creation & ... PTO

& back again. Each Sephiroth or Globe is related to a specific No. from 1 (Heaven) to 10 (Hell) & attributed to specific m.Sqs. Refer to pp 61 + 226. Qaballahists call all ten collectively the MERKABAH or "Chariot" of God, whereby He/She could descend from Heaven into men's/women's Souls. See also MERKABAH & QABALLAH.

SEPHIROTH – IN KABALISTIC TERMS, SPHERES WHICH REPRESENT STAGES ON THE WAY FROM NOUMENAL EXISTENCE TO THE BUILDING OF A PHYSICAL VEHICLE IN THE PHENOMENAL WORLD. THE TEN LIGHTS REFERRED TO IN THE SEFER YETSIRAH (BOOK OF SPLENDOR) WHICH FORM THE TREE OF LIFE.

SEQUENTIALLY: CONSECUTIVE

SEX: (by Kahlil Gibran):

Sex

The most highly sexed beings upon the planet are the creators, the poets, sculptors, painters, musicians . . . and so it has been from the beginning. And among them sex is a beautiful and exalted gift. Sex is always beautiful, and it is always shy.

—MS-94-95

SEXTILE: Latin "sex" = six. 60° angular position of two heavenly bodies. see TETRAHEDRON. Our original calendar was based on 360 days per year before the tilt of the Earth's axis. Thus a Circle or annual revolution of Earth around the Sun was termed 360°. A sixth of this is 60°.

SHABDA: Skt. (or Shabd Nahd) Absolute inner sound. Cosmic supersonic current of vibrational energy manifested as unstruck sound; or psychic white noise or background note of consciousness. Can be heard internally in meditation & created externally by a Siddha yogi with chanting (Graphic from 'Auralgan' Ear Drops, Whitehall Laboratories).

SHAKTI: Skt. the feminine creative, kinetic principle of existence, often personified. She is Power, energy permeating creation. The transmission of Energy is known as SHAKTIPAT. Shown here is Shakti's Magic Square where all pairs in the petals sum to 10 & all diagonals = 20. See also BISA YANTRA. Sanskrit numerals shown:

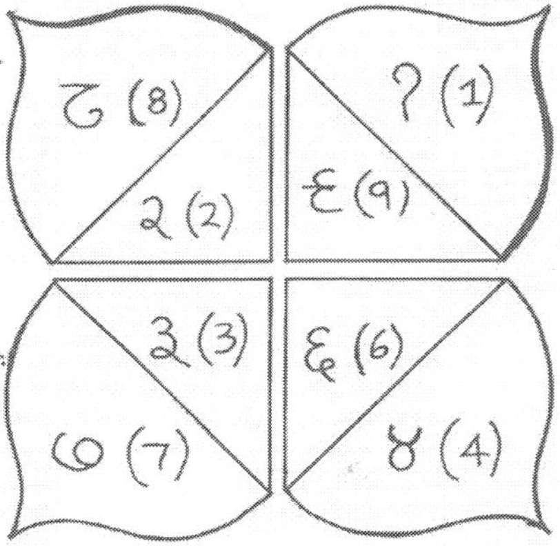

SHAN-MUKHI-MUDRA: Skt. This exercise of closing 6 doors leads to The Opening of The Tenth Door i.e: seals the 9 bodily

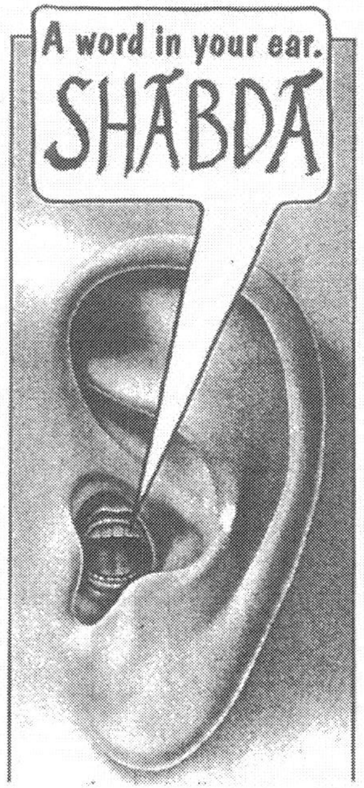

shan-mukhi-mudrā ("six-openings seal") is referred to, for instance, in the *Goraksha Paddhati* (II.16), where it is said to consist in the blocking of one's ears, eyes, and nostrils with one's fingers. This "seal" (*mudrā*) is correctly executed by covering the ears with one's thumbs, the eyes with one's index fingers, and the nostril with the remaining fingers. This practice is recommended for the manifestation of the inner sound (*nāda*).

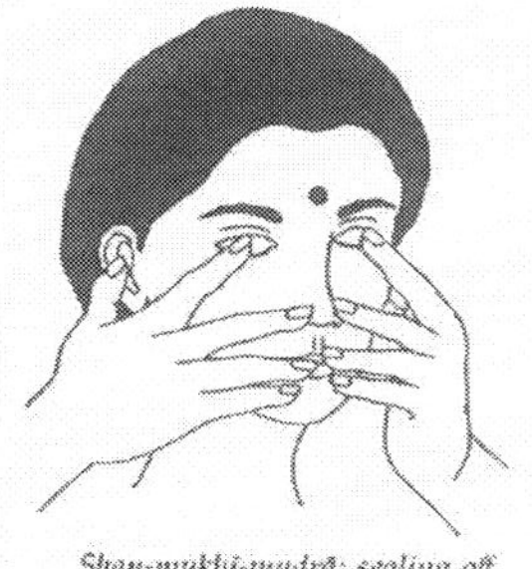

Shan-mukhi-mudrā: sealing off the six openings of the head.

orifices: 2 eyes, 2 ears, 2 nostrils, 1 mouth, 1 anus & 1 penis/vagina. The latter two are locked (bandha or constriction) such that the 10th Door or 3rd Eye region is illuminated like an internal computer video holographic screen showing rapid successions of the etheric geometries

SHANTI: Skt. Peace

SHIVA: Skt. lit. "the Benevblent" personified form of the Divine, yet paradoxically the Destroyer of the Universe. The dancing destroyer, forms passing, forms evanescent. Essentially, Shiva destroys the ego personality so that it is pervious to the divine light. As Lord of the Dance, the "ego" is squashed underneath his foot. His consort is SHAKTI who has 9 aspects such as Kali, Durga etc.

SHRI YANTRA: Skt. lit. "blessed device", or Sri-Chakra, has 9 interlacing triangles representing the feminine (the 5 downward Δs) & the masculine (the 4 pointing upwards). They yield a total of 43 small Δs relating to particular aspects of existence. Sanskrit letters or seed sounds (bija–MANTRAS) are inscribed as names of the deities. All this is geometrically enclosed by lotus petals, concentric rings & a final outer protective square known as the "world house". This is the most famous of all Indian power diagrams. see pp 92, 93, 96 – 100.

SIFR: Ab. zero. whence cipher.

SIGMA: Gk. letter symbol of the alphabet, used for: The Sum of the Terms.

$$\sum_{1}^{4} = 10$$

eg: The sum of the first four Nos. is written as: shown above.
eG: Sigma 36 = 666. see also P 127.

SPHINX : The symbolic vehicle of Light built in Egypt to symbolize the vehicle that can go through the solar spectrum. The vehicle has the "face" of higher evolutionary intelligence which is capable of riding the solar waves (i.e: the Lion = solar travel). The "Lion of Judah" in stone as a witness of Man's destiny in space (K.O.E). The Sphinx also suggests an advanced race from this planet or beyond that has mastered genetics & the hybridization of the species. Half Lion, half Women relates to the cusp of Virgo (Woman) & Leo (Lion). It is thought the Sphinx faces the constellation of Leo. Secret tunnels are rumoured beneath the Sphinx. It may also link up with the largest tunnel in the interior of this Earth & that is from Ayer's Rock to Tibet!

Sphinx. A fabulous being composed of several parts of the human being and four of various animals. The sphinx at Thebes had the head and breasts of a woman, the body of a bull or dog, the claws of a lion, the tail of a dragon and the wings of a bird (8). Being the supreme embodiment of the enigma, the sphinx keeps watch over an ultimate meaning which must remain for ever beyond the understanding of man. Jung sees in it a synthesis of the 'Terrible Mother', a symbol which has left its mark on mythology as well (31). The mask of the sphinx pertains to the mother-image and also to nature-symbolism; but beneath the mask lie the implications of the myth of multiplicity or of the enigmatic fragmentation of the cosmos. According to esoteric tradition, the Gizeh sphinx is a synthesis of all the science of the past. It is shown contemplating the rising sun and seems to embrace both heaven and earth in its meaning. It is, of course, a symbol which unites, in the midst of the heterogeneity of existence, the four Elements (corresponding to the tetramorphs) with the quintessence or the spirit

SPIRAL : A schematic image of the evolution of the universe. It is also a classical form symbolizing the orbit of the moon, & a symbol for growth, related to the Golden Numbers of the FIBONACCI sequence : 1 1, 2, 3, 5, 8, 13, 21, 34, 55, 89, 144 etc.

SQUARING of the CIRCLE : The Mystical Quadrature of the Circle has the task of making the perimeter of a Square (= Earth) equal to the circumference of a Circle (= Spirit). The triangle obtained through this construction has the same angles as the Pyramid, 51° 51' at base (see dotted lines). Refer to pp 122 & 123. Also see KAABA on p 263 & VESICA PISCIS.

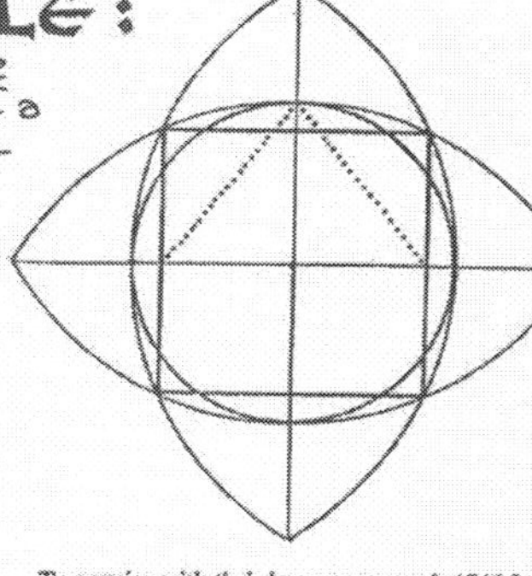

Two vesica with their longer axes, each 1746 feet in length, placed at right angles, enclose a circle with diameter 1008 and circumference 3168, the dimensions of the Holy City. This figure is the perfect symbol of the cosmic temple as equilibrator of forces.

STAR TETRAHEDRON : see MERKABAH

STEINER : Rudolf, designed a spiritual & creative school curriculum in Germany, now world wide, based on seasonal & 7yr cyclic principles e.g: a child has their teacher for 7 years, the child begins learning their AB & 1, 2, 3 at the age of 7. His followers are called AnthroPosoPhists. The above diagram is an example of the sacred Geometry that is taught.

SUFI : Islamic sect who honour all religions equal. Known as whirling Dervishes, they spin for hours & drop in ecstasy as worship of the Absolute. Today, a Sufi is anyone who studies Comparative Religion & can enter a X'tian church or Moslem mosque. The Sufi Magic Squares have No. 1 at their centre, symbolizing the Heart as the End & Beginning of the Divine Journey. see p 72. see also p 262 'Hermes'.

SUNYA or SUNYATA: Skt: Empty of Being, or Non-being, in a de-personalized awareness of the true nature of spaceless consciousness. Unqualified emptiness and purity of consciousness which radiates. Sunyata is the formlessness at the basis of all form, just as silence is at the basis of sound. Of such is the highest knowledge—PRAJNA.

SUNYA :

SUSHUMNA: Skt: A psychic channel of the bio-electrical field of the human entity in the center of the spinal column, related to the central nervous system and brain function. Part of the "Tree of Life" referred to in Genesis, guarded by two angels. viz: The two ends of the Shakti/Shiva Nadis in the nasal sinuses.

see CHAKRA p 157.

SWADISTHANA: Skt. The second psychic center in the region of the spleen above the genitals. It radiates psychic force in six sections. Controls our relationships to our fellow man and society and our response to cultural patterns of thought.

see CHAKRA p 157.

SWAMI : or Svami, Skt, one who can control his mind & senses, lust & greed, & transcends blame or victim consciousness.

SWAZTIKA : or Svastika. Skt. "Su" = "well" & "asti" = it is ∴ means auspicious or fortunate. It is a sun symbol in that the m.Sq. of 6x6 has its sum of the 36 digits adding to 666 (see SIGMA p 272). Its m.Sq. Pattern or YANTRA creates the Swaztika. (Refer to Vol 1, Ch 6) & it relates to the 6th SEPHIROTH on the Tree of Life. Hindu builders always bless a house by placing a swaztika under the first cornerstone at the onset of building, with priests & chants & PUJA.

SWARA SADHANA: The practice of a yoga of rejuvenation which reverses the normal breath and causes the shakti force to flow through the left nostril from sunrise to sunset; by breathing through the right nostril from sunset to sunrise the Tantric student or yogi causes the Shiva force to flow. Thus Shakti (moon breath) is made to flow while the sun is shining and the Shiva (sun breath) flows during the moon's period, predominantly when it is waxing, until the ageing process is reversed.

SYMBOL : The m.Sq. morphs (forms) or symbols hooks us into the Source of all knowledge withIN our cells. José Argüelles, in 'The Mayan Factor' explains : cont:

"... A symbol is a resonant structure of a particular quality of radiant energy that takes form in our senses. Like different kinds of radar stations, the senses continuously receive the inrush of resonant waveforms that comprise our universe. Plato & Jung called these resonant structures "archetypes", form-constants that exist in & define a field of consciousness that transcends both time & the individual. As vibratory form-fields, symbols literally create, work with & inform our Light Body. The Light Body is the electro-resonant galactic code-bank that informs the genetic code-bank ..."

SYNERGY: Word coined by Buckminster Fuller, derived from the Greek "Synergos" means "working together". The Law of Synergy appears in the Bible in Matthew 18:20 "For where two or three are gathered together in my name There Am I in the midst of Thee..." Synergy = Synthesis + Energy. Synergy is the divine key for the geometric expansion of consciousness. Remember the ancient axiom: "God Geometrizes".

TANTRA: Skt. lit. "Loom". Synonymous with "shastra" or scripture textbook, the term refers to sacred writings on psycho-cosmology, the history of the world i.e: yugas or ages, worship of deities & of rites for the acquisition of a battery of magical powers & esoteric processes like KUNDALINI-SHAKTI: Tantric adepts affirm the body as the temple of the Divine & sexual practices are focussed on a man not losing one drop of sperm. see BINDU p 258. Tantrism makes use of sexual symbolism to express ultimate polarities. The decapitated Goddess Chinnamastā represents the spiritual law of self-sacrifice.

TANTRA YANTRA

The Sun surrounded by three crescent moons, symbolizing the unity of male and female aspects of the psyche, of the male-firm-active Yab/Yang and the female-receptive-yielding Yum/Yin principles of the universe.

The word Tantra is related to the concept of weaving (thread, web, fabric, etc.) hinting at the interwovenness of things and actions—the interdependence of all that exists, cause and effect continually interacting. Tantra is the mystical path of knowledge whereby all experiences in the world are seen as material for the burning of Karma and the gaining of Enlightenment.

TAT TWAM ASI: Skt. That Thou Art.

TESSELLATE: in regards to a specific m.Sq. to tile or place a certain YANTRA side-by-side, above & below, to form large fractal / recursive patterns, in simple chequered or mosaic forms.

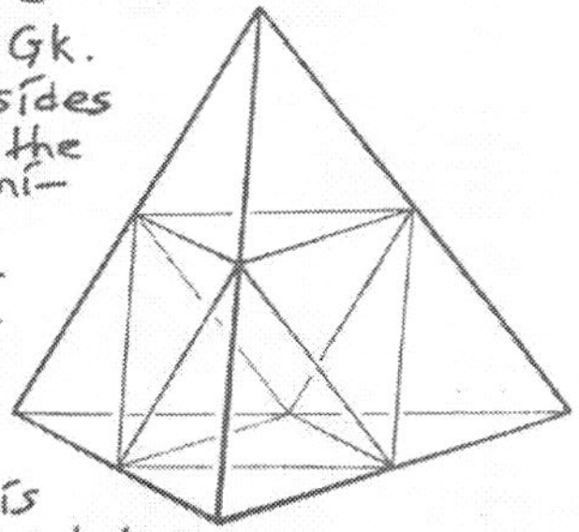

TETRAHEDRON: Gk. "tetra" = 4, "hedron" = bases, sides ∴ = 4 sided pyramid. It is the basic building block of the universe + atomic structure. A triangle is the only structurally stable polygon. According to Buckminster Fuller, the tetrahedron is the minimum omni-triangulated omni-structural system. It is the most profound & fundamental of all structures. eg: two interlaced tetrahedra (A 3-Dimensional version of the ✡ shown here or Star of Solomon) forms the Star Tetrahedron whose vertices, when joined, form a Cube which is the map of the biblical New Jerusalem. Nature is most economical ∴ she could not use cubes to quantitate her structurings. The cubes, as you know, represent our x, y, z axes / co-ordinate system, based on 90° angles, whereas the 60° angles of the Tetrahedron form the basis of ancient cosmic SEXTILE grids e.g: Babylonian counting & calendrical systems. Notice how, in the most top diagram an Octahedron is formed inside the Tetrahedron, when the 6 midpoints of the 6 edges are joined. Thus the Octahedron & Tetrahedron are known as DUALS. The other pair of Platonic Solids that are DUALS are the Icosahedron + Dodecahedron.

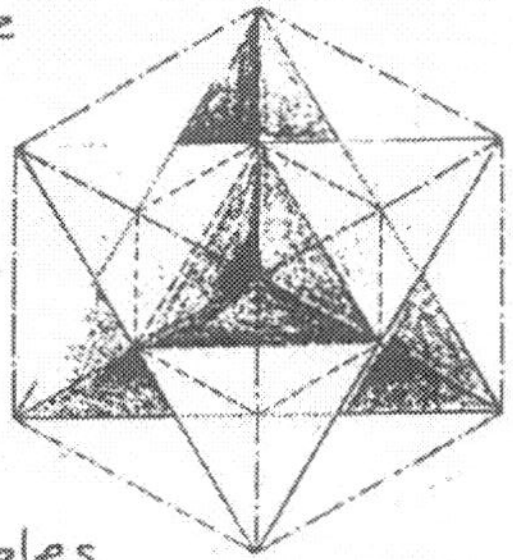

TETRACTYS: Gk. "tetra" = 4. A triangle of 10 Dots (1 + 2 + 3 + 4 = 10) that was Pythagoras' symbol of great importance, revealing the ...

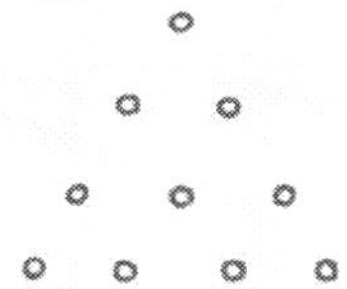

cont.

mysteries of the universe e.g: the Platonic Solids & ∴ atomic structure are hidden within the Tetractys. Pythagoras, circa C6th BC was vegetarian & believed in the Divinity of Number Vibration, in the Transmigration of Souls, & in the Holiness of the number Ten. SIGMA 1 to 4 = 10. See page 272. Sometimes spelt as Tetraktys.

$$\sum_{1}^{4} = 10$$

TETRAGRAMMATON: Heb. the unpronouncable 4-lettered holy Name of God see p 167. Yod – He – Vau – He. יהוה

THALAMUS: the repository of images in your brain.

THEOLOGUE: Lat. "Theo" = God, "Logue" = discussion ∴ a talking with The God I Am, the God You Are.

THEOSIS: The drawing in of All into one's consciousness. The imbuing of the Being

THEOSOPHIA: Gk. "Theos" = God, "Sophia" = Wisdom. The modern word is "Theosophy" = doctrine of insight into the divine nature or revelation p 205.

THERAPEUTAE: in Judaea & Syria, holy visionaries at the time of Iesus, were called Essenes. The Essene BrotherSisterhood that lived by Lake Mareotis near Alexandria were known as Therapeutae or healers. They ate of fruits & seeds; played melodic music & grew very old. Their Initiates faithfully took care of books & hid them from the hands of corrupters. Blessed Be.

THIRD EYE: or Ajna in Sanskrit is associated with the pineal gland in the centre of the head. Its location can be plotted by imagining two skewers, one passing through your two ears & the other betwixt the midpoint of your eyes. Its awakening leads to cosmic consciousness. As it develops it joins with the chakras directly above it as a stem & flower. cont.

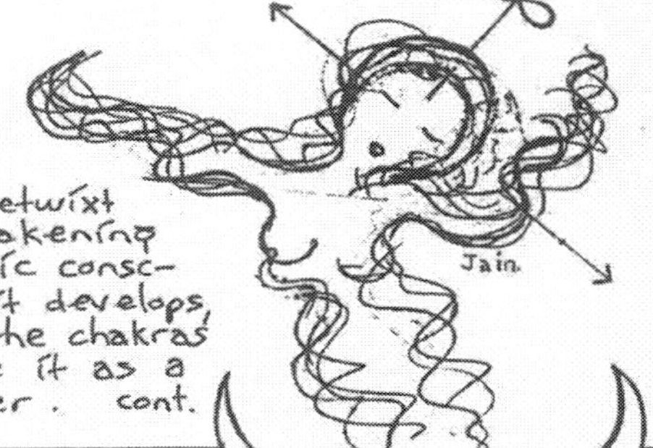

It has a lens, like your physical eyes, with its eyeball facing heavenword. It is the chakra connected with the intellect & the receiving of information. It also regulates the amount of PRANA coming in via the Fontanelle (which is the hole in baby's crowns). Your Third Eye use to be walnut-sized, at present it is an atrophied pea-size, but in many people it is beginning to swell. As the brain increases in size, one may experience headaches ∴ drink lots of pure water to help assist the building of your new Light bodies whose DNA filaments have 12, not 2 helixes ∴ you are acquiring your ancient full faculties ∴ drink lots of wheatgrass juice & consume silica as found in outer leaves of organic lettuce. Grow your gardens. You are replacing carbon-based bodies with those of crystal or silica. Silica's atomic structure is an infinite array of TESSELATED STAR TETRAHEDRONS (i.e tiled three dimensional Star of Solomons ✡✡✡✡✡✡). The Third Eye represents your Spiritual Vision & focus. p 169.

TRANSPOSITION, METHOD OF: Beginning with a NATURAL SQUARE, & like origami, simple triangular folds & steps to move numbers are made to create the M. Sq. Works only for odd Numbered M. Sqs. i.e 3, 5, 7, 9, etc The word "transpose" lit. means "to move across". see pp 68 – 71.

TRATAKA: As an aid to concentration, to stare or gaze mindlessly without blinking, at an object like a candle flame or YANTRA, until tears begin to flow. It not only develops the THIRD EYE, & leads to ClairVoyance, but helps cure eye afflictions. see also the Tibetan Eye Exercise on p 50, fig 35 which strengthens the eye muscles.

TRIGRAM: see I-CHING on p 263 & p6.

TZOLKIN: is the Mayan Calendar's cyclic harmonics of planets, suns & galaxies. Its a very simple galactic code of 13 numbers (the 13 full moons of each year) & 20 symbols that forms the most precise mathematically-based calendar on : cont...

TZOLKIN AS THE HARMONIC MODULE

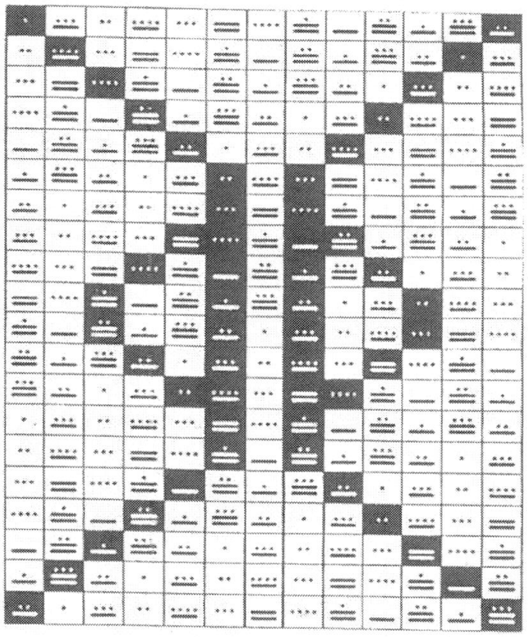

Earth. Now 13 x 20 = 260 = one cycle of time in the Tzolkin's multi-dimensional perspective. Interestingly, 260 is the CONSTANT or Magic Sum of the M. Sq. of 8 x 8. And the M. Sq. of 8x8 relates of course to the 64 HEXAGRAMS of the Chinese I-CHING, which is another cryptic DNA matrix (see p 263). DNA has been shown to be a phenomena of 4 cubed, i.e $4^3 = 64$. This is well explained in José Argüelle's "The Mayan Factor". Hunab Ku is the Spirit or Name of our Galactic Core / Center. The Tzolkin is the Universal Harmonic Matrix & the module for the Psi Bank or the planet's memory field. The Mayan Great Cycle or Hologram of Time is 5,125 Earth years & ends in the year 2012 A.D. on December 21st.

UPANISHADS : Skt. lit. "to sit down close to one's teacher". Ancient teachings of the forest dwellers (RISHIS), referring to a body of spiritual texts in which esoteric knowledge was transmitted by word of mouth from teacher (guru) to disciple. Sacred revelations on wisdom & secret teachings (rahasya) dating as far as Philosophical sections of the VEDAS, numbering 108.

(JNANA) c9th BC.

VAIRAGYA: Skt. the falling away of worldly desires as a result of spiritual Growth & Practices.

Divine apathy or detachment from the fruits of all action. Said to be the highest state of spiritual achievement, akin to total absorption in God-Consciousness, where no difference is seen between the human and any aspect of creation such as grass, insects, stars, etc.

VAISNAVISM : Skt. the Absolute Love binding all men/women despite caste or societal conditionings. A religion of Liberty.

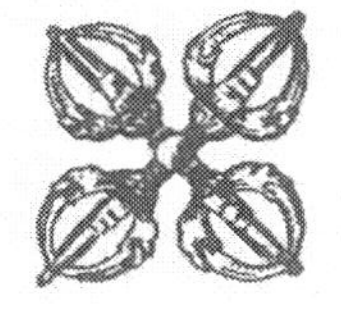

VAJRA : see DOUBLE DORJE p 260.

At the nuclear center of Being we become aware that beyond the world of the senses is the stillness of the most pure brilliant consciousness which the Sanskrit seers called VAJRA. The word VAJRA is difficult for translation and so even in Tibet the great yogis used their own word DORJE.

VAJRA-SATTVA:

Symbolical of the inspiration which distinguishes the real teacher of wisdom from the learned acquisition of knowledge. VAJRA (cosmic consciousness of immortality and indestructible substance) and SATTVA (enlightened direct perception or wisdom). In its active state or dance posture it begins to turn on the world, stamping on the ego. The chopping knife symbolizes cutting through ego-attachment in thoughts of self and the skull bowl symbolizes mastery over death by the skull cap brimming over with bliss.

DORJE-SEMPSA in Tibet has the same meaning. Literally it is the indestructible consciousness of eternal Being (VAJRA) with wisdom and power over cosmic order (SATTVA). Usually portrayed in its active or passive states or holding a VAJRA sceptre, a symbol of the ultimate cosmic power, called a DORJE in Tibetan.

VAJROLI-MUDRA : Skt. the control over the semen (BINDU-SIDDHI). It is the sexual technique of sucking up the female ejaculate (rajas) with the penis (mehana). To practice this, the yogins inserts tubes in his penis, to suck up water, then milk, then honey, then blow it out again. When mastered, he ejaculates his semen (retas) into the vagina & then draws it up again together with the woman's ejaculate (shonita).

VASTU PURUSHA : Skt. lit. Supreme Principle. A mandala of ever-increasing squares, ritually drawn on temple ground floors to consecrate its architechural resonance. Its central square is dedicated to BRAHMA see p 51 fig 36.

VEDAS : Skt. the most ancient of Aryan teachings & hymns. The original literature, comprising all branches of PHILOSOPHY, Science, Art & Spiritual Knowledge, which was first spoken by Lord VISHNU to the creator BRAHMA.

VENTRAL : belly, abdominal e.g. the M. Sq. of 3x3 or Lo-Shu appears on the ventral of the legendary Tortoise see pp 6, 7, 264.

VESICA PISCIS

VESICA PISCIS

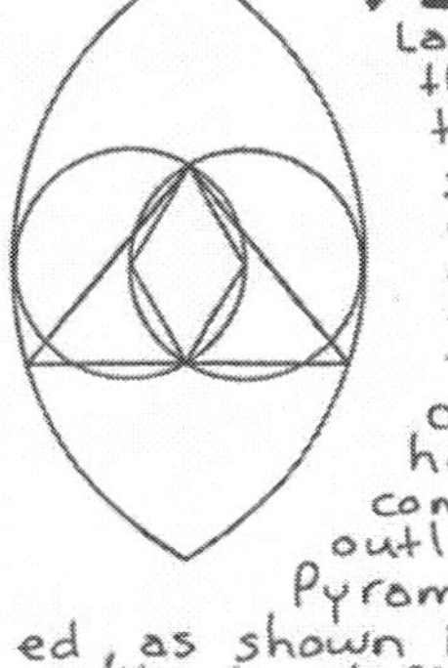

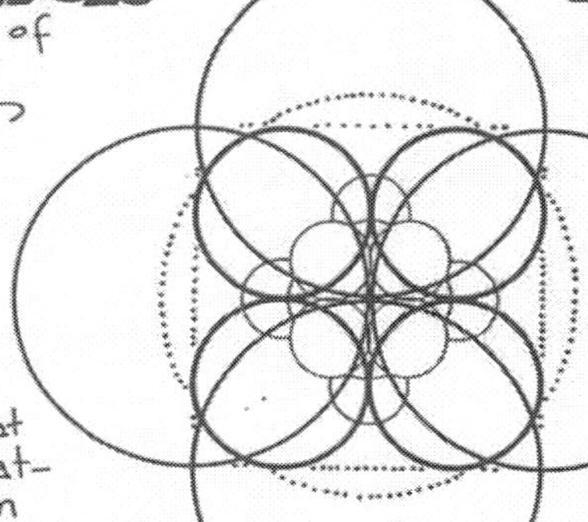

The connecting, cellular spheres producing the primary patterns of order in space.

—see pp 122 + 123.

Latin. Lit. "Vessel of the Fish". It is the significant shape formed when one circle's circumference passes through the centre of another. With mere compass work, the outline of the Great Pyramid can be created, as shown in the diagram on the top left. And in a few steps the Pentagon can be created. Very important to study this in Sacred Geometry. The diagram on the top right is not a true vesica-piscis but notice how the mystical SQUARING OF THE CIRCLE is formed as depicted by the dotted lines. The connecting cellular spheres produce the patterns of order in space (from "Rhythms of Vision" by Lawrence Blair). See PHI, PYRAMID + SQUARING OF THE CIRCLE.

VIBHUTI: Skt. Manifestation. The "far-flung powers" by which Krishna pervades the universe. It is the ashes that the Shaiva ascetics smear on their bodies to indicate their status as SANYASIN or world-renouncers. Sai-Baba can shower vibhuti out of the palms of his hand. He is the most popular of all Indian gurus sometimes attracting as many as 40,000 guests world-wide, from all walks of life.

VIDYA: Skt. lit. "Wisdom". The lore of Sri Vidya required that any yantric artist must purify their energy before commencement of such a holy project. e.g: fasting, celibacy, seclusion. see p 96.

VIPASSANA: Skt. lit. "Seeing things as they are", where attention is turned to constant scrutiny of each successive unit in the thought continuum.

VISHNU: the Preserver of the Hindu trinity. (BRAHMA = creator, SHIVA = Destroyer). Krishna is an incarnation of Vishnu as shown graphically on p 205, in his atomic form & residing in every living Person's heart. p 276.

VISHUDDHA: see also CHAKRA on pp 258, 259 & 157

The Throat center of consciousness which radiates in sixteen different sections. Controls the power of devotion, commitment, ideals, religious dedication, scientific search for concepts, memory, mind, fixation in language and semantic reactions, moral and ethical laws, concept-making faculty and the satisfaction drive to have reality fixed in past time as well as the desire to purify consciousness. Color is a brilliant electric blue, iridescent in its highest manifestation.

WAZIFA: Ab. for "MANTRAM". Unlike ordinary language, where words are linked to their significance by pure convention, the Mantram or Wazifa have been handed down from archaic languages where each sound is the vibratory counterpart of a transcendental significance. On the physical plane every object has a specific rate of vibration which might include several frequencies running parallel.

WRATCHET: to rotate or superimpose or overlay a m. sq. upon itself at any chosen angle.

YAB-YUM: YING-YANG or SHIVA-SHAKTI.

The passive state of the one Cosmic Being, the perfect balance of male & female forces emanating the reality of the clear transparent diamond-hard body of the immortal light from which the soul & the universe are created. The primordial Buddha sitting on a lotus throne in psychic union with the female aspect of wisdom.

YANTRA: Skt. lit. "device, instrument, tool". A geometric representation of the energies of the cosmos & human body (as a microcosmic replica of the macrocosm). Power diagrams that serve to act as "transistor" circuits or cont.

channels conducting psychic essences. Yantras & other symbols are like psychic batteries only better – they are eternal transmitters of spiritual vibes. In tantric ritual, the Yantra must be completely internalized, i.e perfectly visualized. All the M.Sq. Yantras that you are required to draw, are crystalline patterns that need to be programmed with clear & loving thoughts / intentions / focii / resolutions / or wishes. By meditating upon them, singleness of mind or one-pointedness is mastered to prevent aimless wandering or intellectual butterfly hopping. Gazing upon Yantras slays the restless monkey in your mind that constantly swings from branch to branch, subject to subject. A Yantram (singular) symbolizes Order amidst Chaos & Unity beyond Division. These M.Sq Circuits are creational Light Codes. Its the ABC, 123 of Light Technology. They are machines or instruments for moving consciousness.

YANTRA-SIDDHĀIS : Skt. sages to whom the powers of the Yantra have been revealed. see p 95.

YING-YANG :

The union of the female/male forces within one's own Self through meditation brings the great boon of the Great Symbol (MAHAMUDRA). This polarity consciousness of right and left hand energies is seen as universal polarity, enjoying transcendental love and wisdom on all levels of Being. It cannot be portrayed visually but can be meditated on as a sphere of spiralling energies with an attitude of wholeness.

see also p 267 'MITSU TOMOE'.

YOGA : Skt. Union, yoke, linking with the Supreme.

YONI YANTRA : Skt. the Vagina (Yoni) is symbolized by a downward turning triangle that invokes ancestral female power. This "triangle of life" or pubic area of a woman's body corresponds to fertility, motherhood, birth, KUNDALINI force, SHAKTI & the mystery of conception. Notice how Shri Mata AMBAJI'S symbol of the Magik Square on p 256 is a triangle. In ancient BHARAT (India) the whole realm of vibrational sacred geometry + spiritual mathematics was dedicated to the Goddess Energy. See p 93. One of the Magic Papyri equated it with "the 3-cornered Paradise of the Earth".

YONI YANTRA

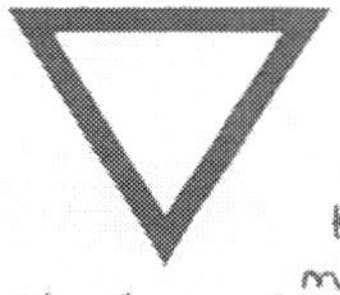

YONI SIGN

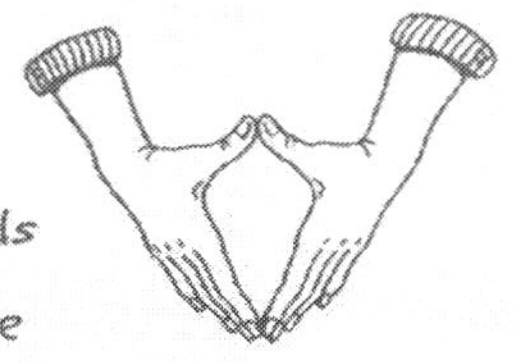

YOGANANDA, Paramahansa (AD. 1893-1952), a master who taught Yoga to the West; set up a centre in 1920 "The Self-Realization Fellowship". He was a pupil of Shri Yukteshwar. Achieved world fame with his 1946 book "Autobiography of a Yogi". This depicted the first published graphic of BabaJi (the deathless Hymalayan Yogi), as shown on p 104. The 3 books that I have up till now published all have been blessed by utilizing this graphic of BabaJi. Yogananda was interested in reconciling Hinduism with Christianity. His poetry uses smart phrases like "I am a Joy billionaire".

One of the most revered spiritual leaders of our time presents a sweeping chronicle of the soul's journey to enlightenment.

YHWH : Heb. Yod Hey Vod Hey of the Everlasting Light. The Unpronouncable Name of God is one of 72 sacred names of the Infinite mind, each having its own universe & celestial hierarchy (K.O.E.). see also TETRAGRAMMATON on p 275 & p 167 & p 246 Ack 167c

ZAT : or DHAT. Skt. That from which all springs; the basis of all phenomena.

ZERO : The Hindus used the word SUNYA meaning zero, & the Arabs translated it to SIFR, meaning empty. Later it became CIPHER & finally in Latin the word was "Zephirum", which later was shortened to ZERO.

ZERO POINT : According to Cliff Pound, Free Energy scientist of Byron Bay North N.S.W., Zero Point Energy is also known as the Ether Field. Ether has no spin, no motion, but has total uniformity. The Ether Field is potential energy. The first distortion is Magnetism. Magnetism is a Spin-Distortion of the Ether Field which is regarded as anisotropic i.e, has uniformity in all directions, whereas "Scalar" has potential but with no direction eg: Heat. Regarding spin: Atomic Spin is the Spin orientation of the magnetic moment of an atom.
If all this sounds too pedantic, don't worry, forget the intellect & tune into your heart, into this last Glossary word:

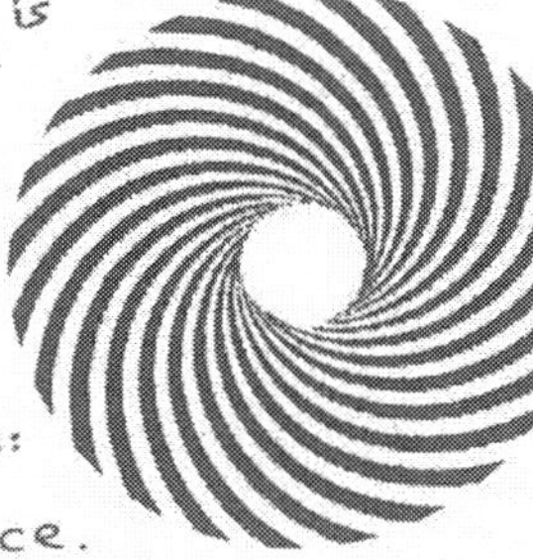

ZIKR : or DHIKR = Remembrance.

INDEX

LEGEND :

m : Sq. = magic
p = page
pp = pages

France, 16th cn.

Page 283

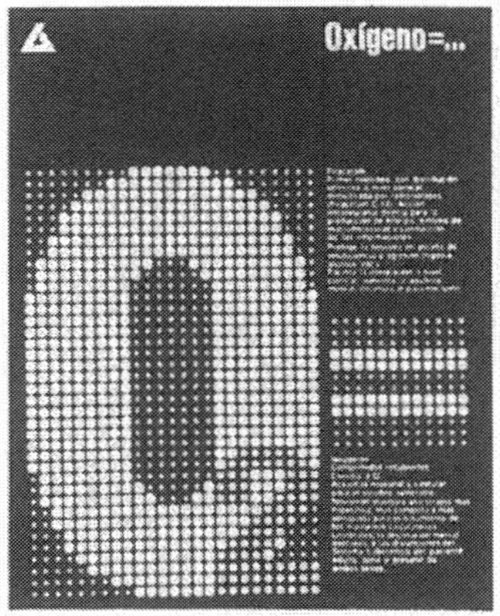

Jain.

The frustrum of a pentagonal pyramid.

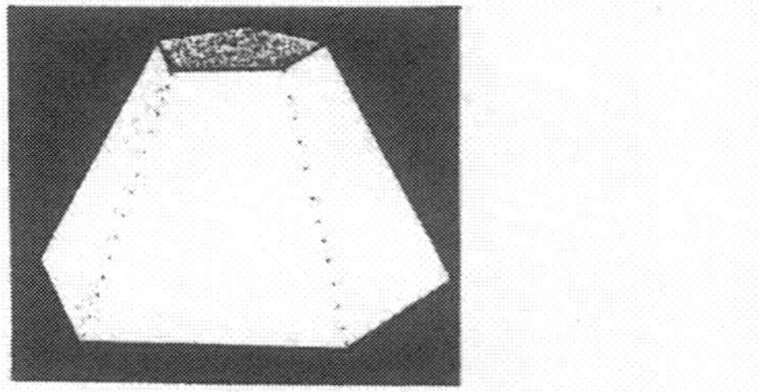

Fold up this figure to make the frustrum of a pentagonal pyramid

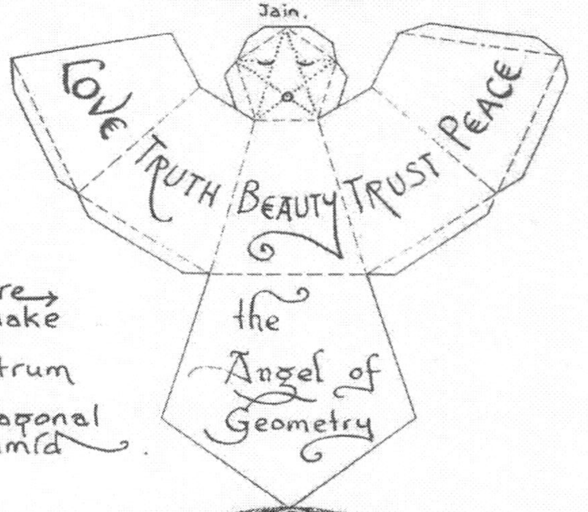

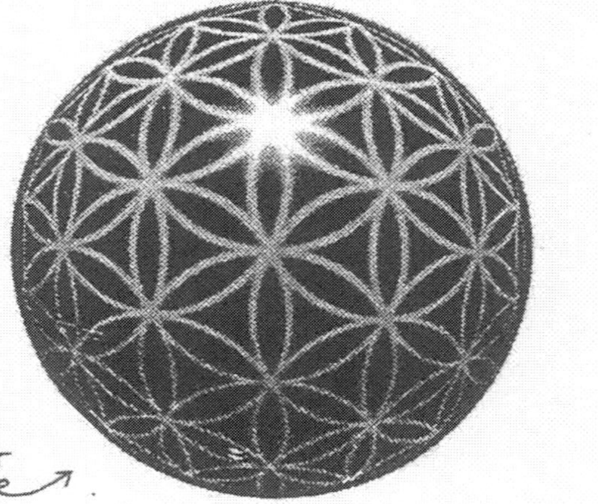

Three Dimensional Flower of Life

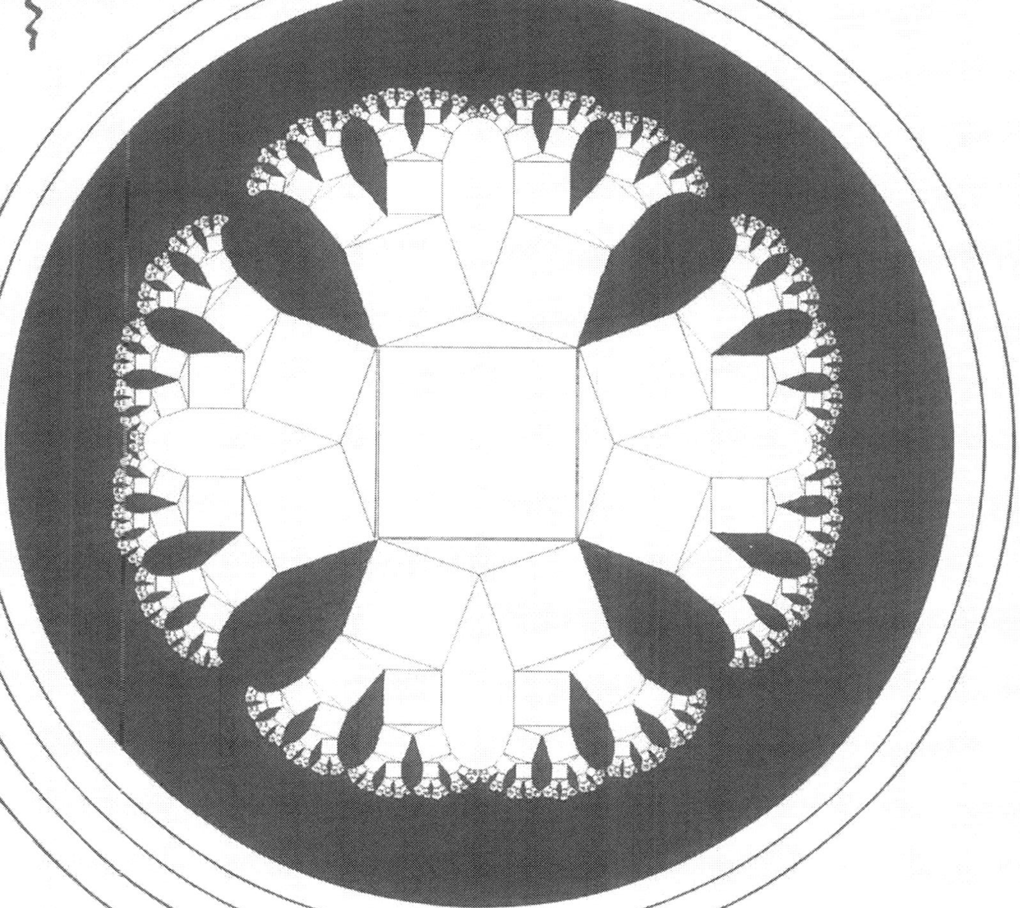

"Towards the end of his career, Gaudi asserted that the straight line belonged to men, the curved one to God".

The time has come, my dear reader, for the scaffolding to be taken down, for the inner structure (the Temple of your Inner Being) is nearly complete.
The Mandala becomes a map not only of our entire universe but also of our inner Soul-Landscape. There is no escape from the PsychoSphere of the Mandala.

Magic is a word of Persian origin, the root being "MAGI", meaning a Persian priest, an enchanter, a wise man who interprets dreams, a fire-worshipper. Related words are 'Magister' = a teacher or master, & 'Magus' = a magician of adept standard. Reference to any good dictionary reveals a scattering of words using the Greek prefix "MAG" meaning great, similar to the Sanskrit "Maha" as in Maharaja (great king). Such words with magical connatations include 'Magistrate' = a ruler, 'Magnate' = a great man, 'Magnificent' = doing great things, 'Magniloquence' = elevated language, 'Magnanimity = Greatness of Mind, & 'magnetism', etc.

Some scholars have interpreted this reverence of M.Sq. studies as a Mathocentric religion. Rather, the message is not to study maths but to pursue your passion & follow your bliss. That is all!

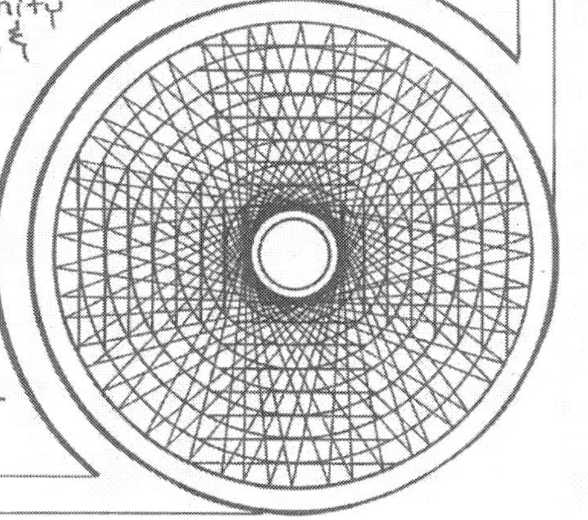

KARL MARX CONTEMPLATING THE ILLUSION OF THE GRAY SPOTS!

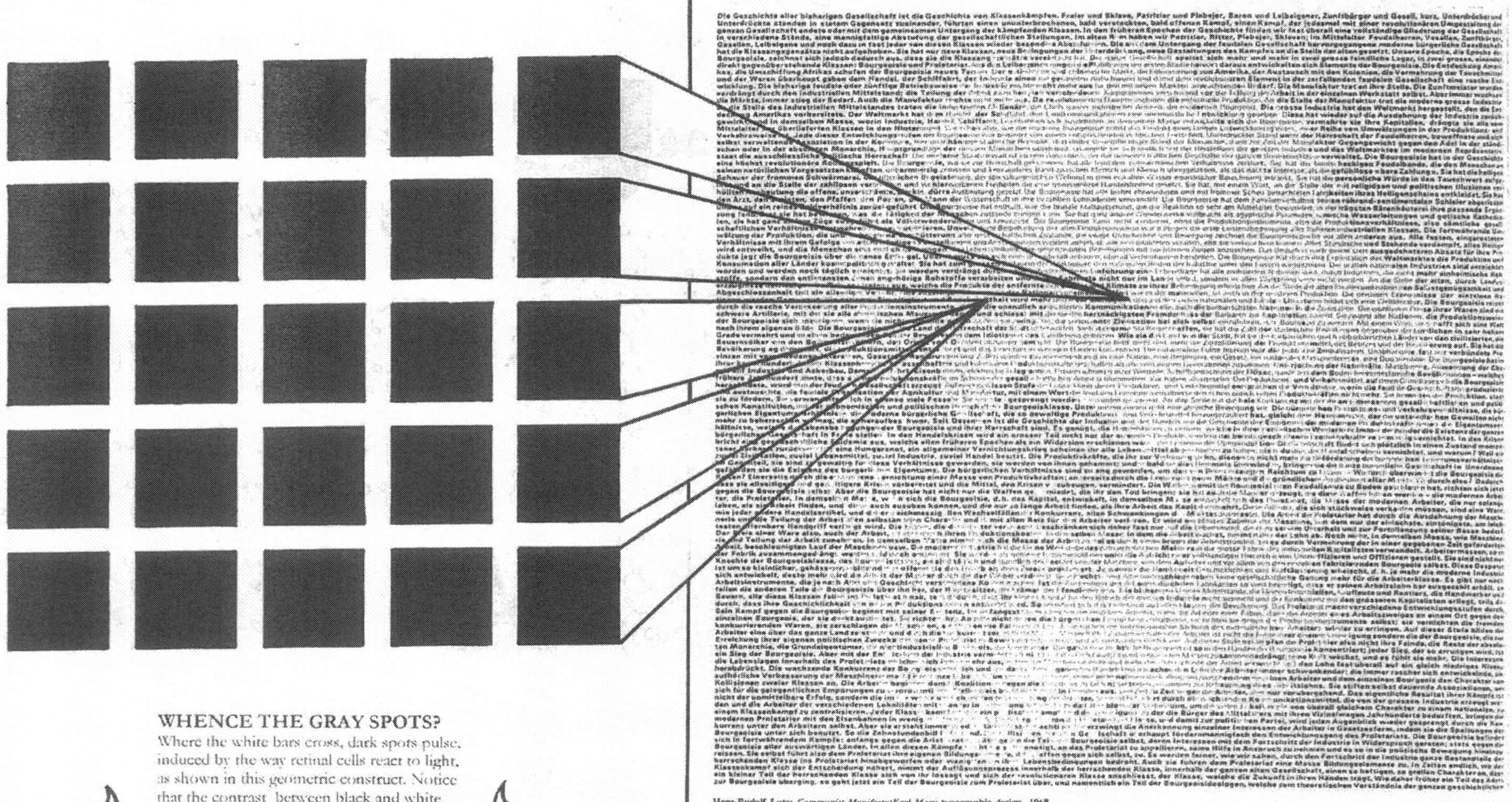

Hans-Rudolf Lutz: *Communist Manifesto/Karl Marx* typographic design. 1968

WHENCE THE GRAY SPOTS?
Where the white bars cross, dark spots pulse, induced by the way retinal cells react to light, as shown in this geometric construct. Notice that the contrast between black and white selectively excites or inhibits the response of these cells, thereby creating the illusion.

CONCLUSION

Nightcap Forest Festival 1992

"NIMBIN DREAMS" Photo by Harsh Prabhu.

Imagine if every word in this Index was a whole set of volumes of books like the modern day Brittanica. Could you now imagine how much knowledge your brain cells contain. All knowledge of Earth's History, the true unadulterated knowledge is all in there, in your own internal bio-computer silicon-chips; every experience of every event, every life, the totality of your numerous emotional sagas & memories, are in there; millions & billions of diaries filed away; all data of the ancient Sciences, Mathematics, physics etc is all in there, locked away in the cells of your body, even in your bones; its all vibrationally recorded withIN, for you are The SOURCE, you are The ALL THAT IS, but somewhere in your Earth's History, manipulators visited, Genetic Masters came into your Sphere of Consciousness & out of ignorance, fear or lack of Love, they tampered with your DNA Geometries to perfect & heal & rejuvenate their own Star Tetrahedronic matrix. Earth, to Them, is a living biological Library of DNA, a grand experiment that went wrong. Glimpses of some of the Activational Geometries in this book will trigger rememberance of the nightmare. So its like, all the million books of your Divine Mind were scrambled, someone walked in, uninvited, gatecrashed your Library & threw every book onto the ground. Pheww! Could you imagine the mess, the out-of-orderliness. And Day-by Day, Year by Year, Lifetime by Lifetime, you've been organizing your books back onto their correct shelves, sometimes asking "Why, Who Am I? Where am I going? What-ever Happened?" slowly remembering, like an uncoiling snake in the glorious sunshine, remembering the Dream of who you really are. Deep down you know that the Truth is always revealed & crime does not pay. Soon you won't even need books. Your children will be able to place their hand on the front cover of any book & vibrationally read or absorb its contents, or the codes that resonant to Truth. Similarly, the modern definition of Healing is about the physics of Spin & Rememberance of our Divine Grandeur. I would like to conclude this Book of Activational Yantras with one Word:

REMEMBER REMEMBER

We call upon the spirit of evolution,
the miraculous force that
inspires rocks and dust
to weave themselves into biology...
Empower and awaken us to creativity.
John Seed

The Ultimate Majik Square.

page 287

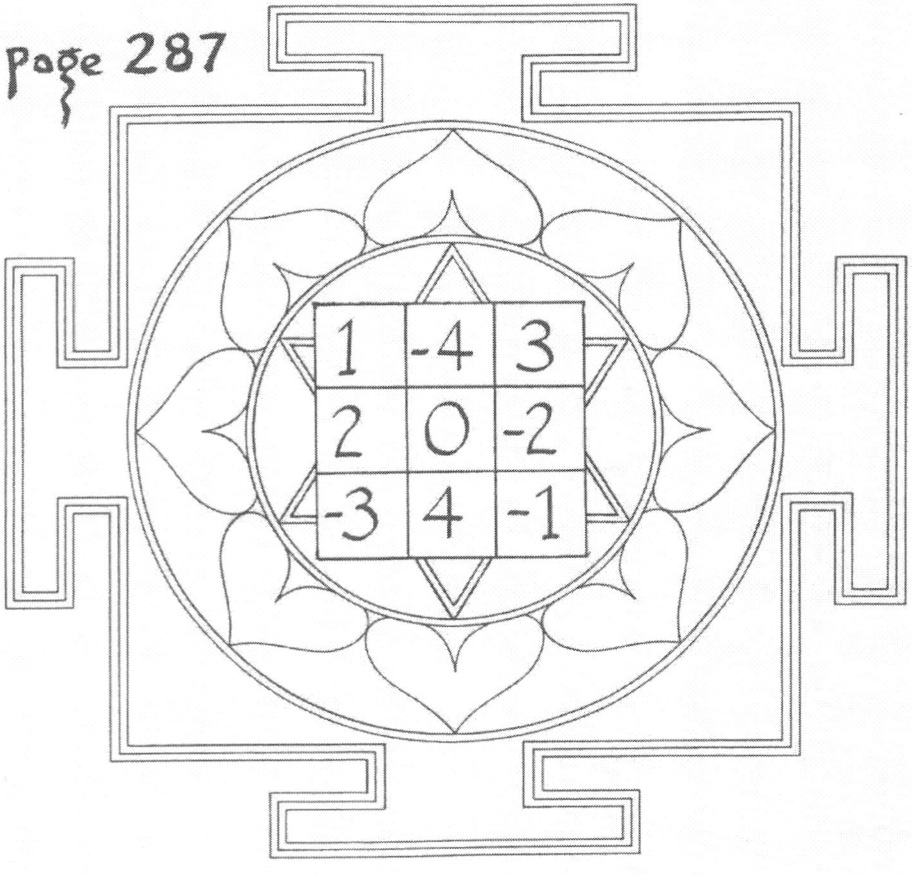

The **CONSTANT = 0** i.e: The Sums of the 3 columns, rows + 2 diagonals all equal 0.

Photo of JAIN by Harmonie Downes.

Resumé

JAIN

ADDRESS :
MAGIC SQUARE STUDIO
50 LEFT BANK RD MULLUMBIMBY CREEK
2482 N.S.W. AUSTRALIA

PHONE: FAX
011 61 2 844409

EDUCATION

SECONDARY EDUCATION

- Kirrawee High School
  Obtained HSC. Top 2%. 1975

TERTIARY EDUCATION

- B.A. Communication
  Institute of Technology, Sydney.

  Electives:
  ~ Professional Writing
  ~ Psychology
  ~ Media Studies
  ~ Film Making

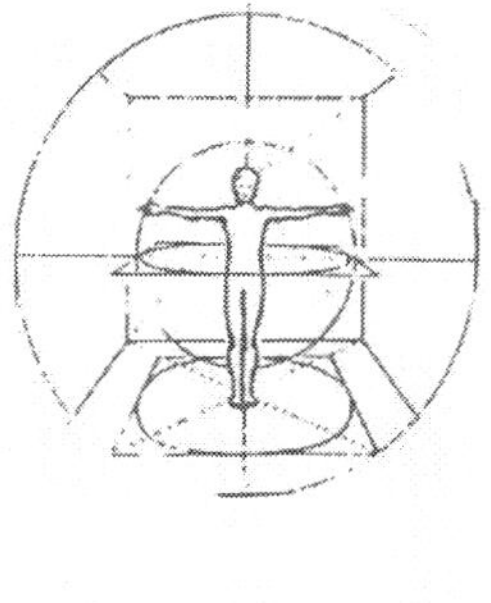

EXHIBITIONS OF MATHEMATICAL ART

Year : 1988
Title : SACRED GEOMETRY
Venue : Redfern Greek Orthodox Church Hall

Year : 1990
Title : THE DIVINE PROPORTION
Venue : Bangalow Hall, North NSW

Year : 1993
Title : THE GATHERING OF THE MATHEMATICIANS
Venue : 'Epicenter' Arts Gallery, Byron Bay

Year : 1995
Title : THE WORLD OF MAGIC SQUARES AND CUBES
Venue : Southern Cross University, North NSW

Year : 1995
Title : THE THEATRE OF THE HOLY NUMBERS
Venue : 'Epicenter' Arts Gallery, Byron Bay

Year : 1997
Title : THE LIFE WORKS OF JAIN
Venue : Magic Square Studio, Sydney

SKILLS

AUTHOR OF 5 BOOKS

1. **THE TIME TWINS 1981**
2. **JOIN-THE-DOTS 1999**
3. **THE BOOK OF MAGIC SQUARES Volume 1 , Yr 2000-**
4. **THE BOOK OF MAGIC SQUARES Volume 2 , Yr 2000**
5. **THE BOOK OF MAGIC SQUARES Volume 3 , Yr 2000**

LOGO DESIGNER.	Designed the political flag for Badu Island, Torres Straits, 1974
DECAL DESIGNER	Designed 3 original geometrical Magic Square patterns on Adhesive Transparencies
LECTURER.	I have lectured to thousands of school children around Australia. I currently run weekend seminars for adults every second month.
THEATRICAL DIRECTOR.	I have developed a Maths Enrichment Program called "Mathemagics" in which Maths is taught through Theatre a.k.a. "The Theatre Of The Holy Numbers".
MURALIST	I have painted 10 large cotton tapestries 3m x 3m Displayed in schools and universities around Australia.

PAGE 260 VOL. 3

SCHOOLS AND UNIVERSITIES WHERE I HAVE TAUGHT SACRED GEOMETRY

1986 ~ Apollo Bay High School, Victoria
~ Montessori, Adelaide
1987 ~ Elands Primary School . Once a week tuition for 3 months.
1991 ~ Kangia Steiner, Murwillumbah
1992 ~ Redcliffe Catholic, Brisbane
1993 ~ Smithfield High, Cairns
~ Cairns West Primary
~ Kadina High, Lismore
~ Ballina High, Ballina
~ Richmond River High, Lismore
~ Mullumbimby Primary, Byron Bay
1994 ~ MATHS CAMP at Tallebudgera Qld
Guest Speaker / Sacred Geometry Seminar
~ Tallebudgera Primary
~ Combabah High, Gold Coast
~ Mount Gravatt High, S.E Qld
1995 ~ Goonellabah Public
~ Ballina High
~ Southern Cross UNIVERSITY, Lismore. Guest Speaker
and Lecturer to Conference of 100 Maths Teachers.
~ Alstonville Primary
~ NORTH COAST MATHS AND SCIENCE CAMP
Lennox Heads
~ Currumbin Valley State School, Qld
~ Grafton High
1996 ~ Bangalow Primary. Passed Exams for School Performance
Validation in NSW Primary Schools.
~ Mount Warning High. Passed Exams for Accreditation in
NSW High Schools.
~ Kangia Steiner School, Murwillumbah
~ Cape Byron Steiner School
~ Currumbin Valley State School
1997 ~ Darlinghurst Primary
~ Mount View High School, Cessnock
~ Korowal School K5-12, Blue Mountains

1998 ~ Lectured on "The DIVINE PROPORTION" for an Adult Education workshop
in Mullumbimby, far North N.S.W. May.
~ Kirrawee High School, June, Sydney
~ St Peters Maths Carnival, August, Sydney
~ Ashcroft High School, August, Sydney
~ Lectured at the National Maths Conference of Australia, with Dan Winter,
with 700 Maths Teachers at the gathering. August.
1999 ~ Bought and renovated my first home / library / studio, known as THE MAGIC
SQUARE STUDIO, located in the rainforest of far North N.S.W. Mullumbimby.
2000 ~ Participated and contributed to the Science-Art Festival / Exhibition in
Murwillumbah N.S.W. and Brisbane. April.
~ Lectured on "The DIVINE PROPORTION" at The Starlight Festival in Bangalow,
N.S.W. April.

Page 261 VOL. 3

PLACES TRAVELLED TO FOR MATHEMATICAL RESEARCH

YEAR	PLACE
1973	Lebanon
1979	Lebanon
1982	Torres Straits
1983	Torres Straits
1986	Australia Tour
1989	India
1990	Tibet
1990	Thailand

REFERENCES

Name: PROFESSOR GARY BELL
Title: President of The Southern Cross Maths Association, Nth NSW University
Address: gbell@scu.edu.au
Telephone: 066 203616
Fax: 221833

Name : RONNIE GAJEWSKI
Title: Maths Teacher, Mount View High School.
Telephone: (049) 911728

Name: ANN JACOBSON
Title: Arts Teacher, Korowal High School
Telephone: (047) 572253

Introducing 2 Magic Sq. Articles.

6	1	8
7	5	3
2	9	4

Fig 1

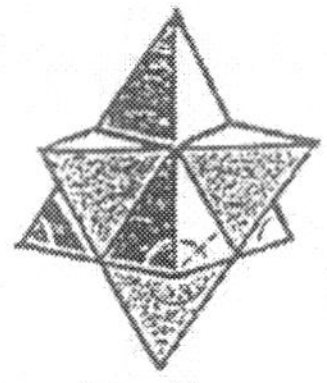

Fig 3

Fig 5

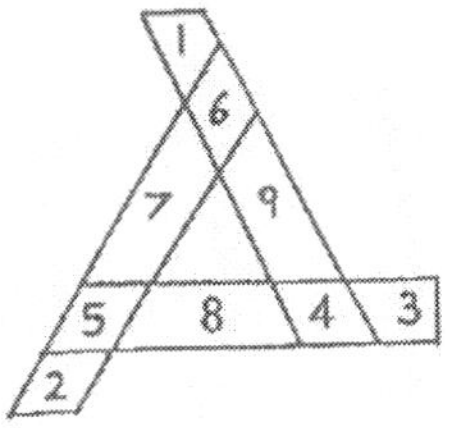

Fig 6

The Magic Square of 3 X 3

A tribute to the Lo-Shu

Drunvalo Melchizedek, in his worldwide "Flower-of-Life" workshops says **Sacred Geometry is a conscious form of Ascension.** He outlines two important generational sequences or eternal codes of systematic growth designs that underlies the organic and mineral kingdoms. They are:

The Binary Code - i.e.: 2 = 1, 2 = 2, 2 = 4, 2 = 8, 2 = 16, 2 = 32, 2 = 64 etc. which is the on-off sequence of computer silicon-chip based systems designed to carry a lot of information. This quickly escalating sequence is a key to the infrastructure of DNA. DNA is a phenomena of 4x4x4 or 4cubed = 64 which relates to the Magic Square of 8x8 whose magic sum is 260 in all directions and corresponds to the Mayan Calendar (Tzolkin) of 260 days and to the 64 hexagrams of the I-Ching.

The second code is the mathematics of Nature expressed by the **Fibonacci Sequence** and one that is advisable for you to remember and learn now: **1, 1, 2, 3, 5, 8, 13, 21, 34, 55, 89.** The next number in this series is merely the sum of the two previous terms i.e.: 55 + 89 = **144.** Each number e.g. 144 divided by its predecessor i.e. 89 arrives at an approximation of 1:1.618.... a transcendental number that keeps going forever. It is known as the Golden Mean Ratio or PHI. Its symbol is Φ a female circle penetrated by a male vertical line. Pythagoras' ancient symbol ☆ has all of its lines intercepted in this ratio. In the human body, the Phi ratio is at the belly button or point of life and fertility. The planetary orbits from the sun are in this ratio and the counter-rotating spiral fields, as seen in pine cones sunflowers etc. obey this sequence.

The purpose of this article is to introduce a 3rd sequence of ancient Fixed Timeless Design known as the **MAGIC SQUARE CONSTANTS.** The mysterious and not well known Formula: **C = n (1+n squared) ÷ 2** determines the magic sum or Constant of a magic square. C = Constant, n = the order or magnitude of the Magic Square. E.g.: This magic square of 3 x 3 from China 5,000 years ago (fig. 1) has an order of 3. Therefore n = 3. By substituting n = 3 into the Formula: C=3/2 (1+3x3) =15. By definition a Magic Square has its columns, rows and diagonals equal in sum. They are this timeless expressions of harmonic order in dimensional chaos or unity within diversity. **In Tibet, every number has a soul, therefore a Magic Square is the epitome of a spiritual community.** Fig 2 shows this third important series of the Magic Square Constants.

Magnitude of the M.S.	Magic Square Constant	Astrological Attribute
n=3	15	Saturn
n=4	34	Jupiter
n=5	65	Mars
n=6	111	Sun
n=7	175	Venus
n=8	260	Mercury
n=9	369	Moon
n=10	505	Earth

Fig. 2

Remember these Constants for they are the keys to the dimensional doorways of increasing frequency. Professor Searl of England has created manned anti-gravity machines (flying saucers or free-energy discs) of which the mathematics of these counter-rotating fields is 100% based on these Magic Squares! Research his work. Magic Squares tap you into the Universal Free Energy Matrix.

Everything has a number or vibration e.g.: Material objects, human organs, colors, sounds, numbers etc. As your Light Bodies are newly being built or restructured you will now need to tap into these codes to kind of 'select the gear' from which you will steer or guide your MerKaBah. Your MerKaBah is your Ascension Vehicle or Spiritual Light Vehicle powered by

continued

interlocked tetrahedronal counter-rotating forces spinning towards the speed of light. Simply, it is viewed as a 3-Dimensional Star of Solomon (fig. 3). Magic Square Artforms can derive ancient symbols like ✡ and 卍 . Before we begin to transform these Magic Square number codes into talismanic art, we need more data.

These 9 Magic Squares listed in fig. 2 correlate to Planetary Energies listed in the ancient Hebraic cosmology known as the Qaballah or Tree of Life. Fig. 2 indicates that the Magic Square of 3 x 3 correlates to Saturn; and the Magic Square of 4 x 4 correlates to Jupiter.

In the ancient Vedas (Indian Scriptures) the Magic Squares of 3 x 3 in their 8 permutated states were combined to form a calendar. Fig. 5 shows how **the Magic Square of 3 x 3** (fig. 1) **is found at the center of the Tibetan calendar/cosmology.** It shows Manjushri incarnating as a fiery tortoise revealing upon its underbelly all that is sacred e.g.: the Zodiac, and the Magic Square of 3 x 3. Hand-sized squared alloy plate of 5 metals has fig. 5 inscribed and it is religiously nailed to door posts outside for spiritual protection of the home.

In India, particularly Pushkar village in Rajasthan where I visited, the Lo-Shu (fig. 1) is framed and placed in shop alcoves where Puja or worship is offered to the Magic Square, that is, offerings of incense and flowers etc. to invoke business success.

Also Yogis wear these numbers engraved on bracelets that fit over the biceps. This is called 'Juntra' and offers the wearer spiritual strength.

The tantric symbol of Goddesses like Shakti and Shri Mata Ambaji is the same Magic Square of 3 but in a feminine triangular form. In fact, the field of Sacred Geometry is always dedicated to the Goddess Energy. The 9 numbers of the Goddess' Magic Triangle (fig. 6) represent the 9 divine spouses of Shiva God. All summations are 20, 20, 20 and is known as the Bisa Yantra (Bis = 20).

The Chinese Book of Changes or "I-Ching" oracles, has its whole philosophy based on this square, where it is known as the Lo-Shu.

It also appears in treatises on Feng Shui which deals with the aesthetics of the ley lines or magnetic energies circulating around our planet.

Many Russian ocean cruisers paint the Lo-Shu on their bow deck, mural-sized, for protection.

Sufi mystics and Hebrew Qaballahists honored a numerologically based alphabet, just like the Greeks, whose language gave a number value for every letter. This ancient study is called Gematria, e.g.: A = 1, B = 2, C = 3 etc. according to this, the Magic Square of 3 Constant, which sums to 15, equated to the unpronounceable Name of the Lord Yahweh, spelt as JAH, a contracted form of the holy Tetragrmmaton IHVH (Yod He Vod He). Then the Magic Square of 4 x 4, whose sums are 34 correspond to the holy name Adonai or Adm or biblically, the First Man. The Magic Square of the Sun has all its sums adding to 111. Therefore the sum (or Sigma) of all its digits from 1 to 36 naturally adds up to the controversial 666. When this Magic Square of the Sun is transformed into art, it produces the Swaztika or Hindu peace symbol, the most beautiful symbol of 'balance in motion' like the spirallic arms of swirling galaxies. (This is fully explained in *"The Book of Harmony Squares" Vol. 1* 1990. Jain)

HOW TO TRANSFORM MAGIC SQUARES INTO ART

Firstly, in fig. 1, visualize a dot in the center of every cell or square. Take away the frame and you are left with a Grid or Lattice of 9 Dots (fig. 7).

Next, connect a line in sequential or consecutive order i.e.: From 1 to 2 to 3 etc. to the last number 9 as shown in fig. 8. This represents the path of Order amidst Chaos.

Then draw a Final Connecting Line i.e.: From 9 to 1. This serves to close the circuit, if you visualize the Pattern or Yantra as a circuit-board of electrical energies, ever-flowing unbroken as shown in fig. 9. The largest Triangles are also in the Fibonacci Sequence which means that these Golden Triangles are pleasing or aesthetic to the eye.

At this moment, your Thoughts are very important as the resultant pattern will be crystalline. Just like any crystal can be specifically programmed.

fig. 7

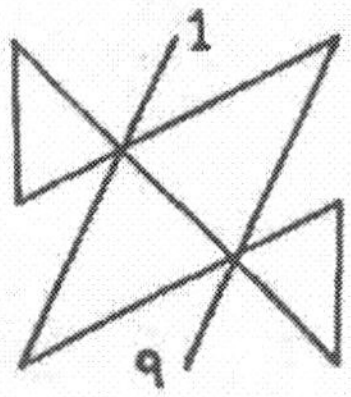

fig. 8

fig. 9

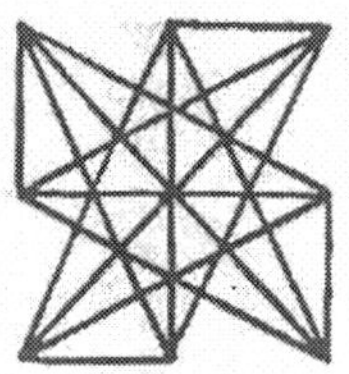

fig. 10

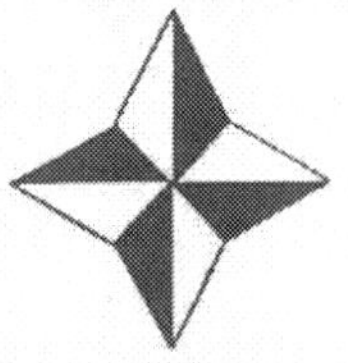

fig. 10a

mentally direct your affirmation into this sacred art or Yantra. A Yantra, in Sanskrit, is the actual 2-Dimensional artform, incorrectly called 'mandala'. The Yantra is the visible pattern that acts as a transistor circuit to conduct psychic energies. The 'mandala' is its 3-Dimensional counterpart or hologram. The 'invisible' mandala that we meditate upon is made 'visible' by the Yantra. Fig 9, in Magic Square jargon, is called "The Lo-Shu or the Magic Square of 3 x 3 at Zero Degree."

By turning the paper around and super-imposing the same image (fig. 9) upon the star-shape a more symmetrical Yantra is formed. It is technically known as 'The Magic Square of 3x3 at 0° +90° '(fig. 10).

The star shape isolated from this Yantra (fig. 10a) when painted gold is the modern metaphysical symbol for the 9th Chakra or the Planetary Gateway according to a 12 Chakra system expounded by Barbara Marciniak, a spokesperson for the Pleiadian mysteries.

By super-imposing the square of fig. 10 upon itself again, according to the Grid-of-Dots format of fig. 11, another more stellated diagram is created from this precise wratcheting effect (fig. 12). This is known as "The Magic Square of 3 x 3 at 4 x 45 degrees". It has become my personal signature for the worldwide transmission of these harmonics and is the logo for "F.R.E.E.D.O.M.S." the institute or mystery school "For Research Expressing Essential Data of Magic Squares".

When fig. 12 is superimposed upon itself, according to another rotation like fig. 11, a more spectacular Yantra appears (fig. 13. This is formed from the original diagram of fig. 9 but ratcheted 8 times at angles of ½ of 45 degrees. This is called "The Magic Square of 3 x 3 at 8 x 22.5 degrees." Such art as generated here is ideal for creative graphic artists.

Imagine this design of fig. 13 as an accurate copper wire circuit-board whose flow of electrons would be so harmonious that there would exist little friction or collision of particles or heat, that the system's energy would actually double in its output. This design of fig. 13 is almost identical to such Rodlin Coil designs as shown in a 1996 Nexus magazine article and reprinted as fig. 14.

So you see, this simple Magic Square of 3 holds within its seed, much hidden knowledge. Many Magic Squares have been emphasized in the biblical Dead Sea Scrolls relating them to parables on the Divine Kingdom. Our modern bible omits these Magic Squares. Why? Has there been a conspiracy to withhold certain knowledge to the masses that may trigger or awaken or assist them to REMEMBER their divinity? Therefore your assignment for The Revelation of the Spiritual Mathemagics is to mentally repeat the cosmic Lo-Shu phone number: 6 1 8 7 5 3 2 9 4. Say it and say it over again.

To conclude, these simple diagrams are what I offer primary and high school students to carefully draw. These **activational** designs are presented to young and old to trigger genetic **REMEMBERANCE** of who we are and which Star Laboratory we came from. I believe that one day the next generation of children will be able to place their hand upon any book and vibrationally "read" or absorb its contents. For such a book "The Book of Harmony Squares" Vol. 1 1990 is available. It contains scores of unique and original Magic Squares to draw and color in. The sequel to this book "The Book of Magic Squares" Vol. 2, 1997 is ready to print, is published, can be ordered. Full of Sacred Geometry material never published before it awaits an eager publisher or sponsor to tune into these codes and transmit them globally.

In love and service
JAIN

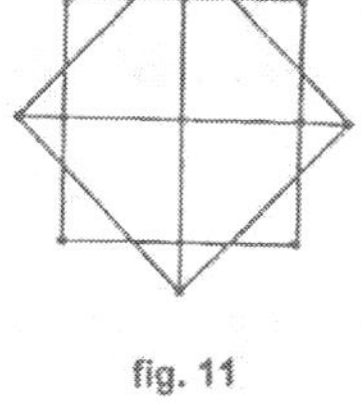

fig. 11

fig. 12

fig. 13

fig. 14

LEAP and the net appears!

This Article was submitted to "EAGLES WINGS" magazine 1996 Vol 4. No 6.

THE PARABLE OF THE SEVEN PALMS

A Treatise on the Venusian Magic Square of 7x7
as gathered by Jain

“...and Iesus, when he came to a certain place where seven palm trees grew, gathered his disciples around him, and to each he gave a number and a name which he only knew who received it. And he said unto them, stand ye as pillars in the House of God and show forth the order according to your numbers which ye have received.

And they stood around him, & they made a body four square, & they said unto him, Lord we cannot. And Iesus said, let him who is greatest among you be even as the least, and the symbol of that which is first be as the symbol of that which is last.

And they did so, & in every way was there equality and yet each bore a different number, and the one side was as the other, and the upper was as the lower, and the inner as the outer. And the Lord said, it is enough. Such is the House of the Wise Master Builder. Four Square it is, and perfect. Many are the Chambers, but the House is One.

Again consider the Body of man, which is a Temple of the Spirit. For the body is one united to its head, which with it is one body. And it has may members, yet all are one body and the one Spirit ruleth and worketh in all; so also is the Kingdom.”

The Parable of the Seven Palms

22	47	16	41	10	35	4
5	23	48	17	42	11	29
30	6	24	49	18	36	12
13	31	7	25	43	19	37
38	14	32	1	26	44	20
21	39	8	33	2	27	45
46	15	40	9	34	3	28

fig. 1

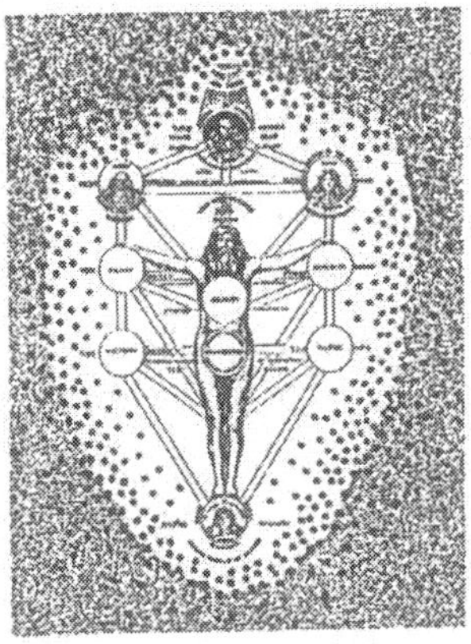

fig. 2

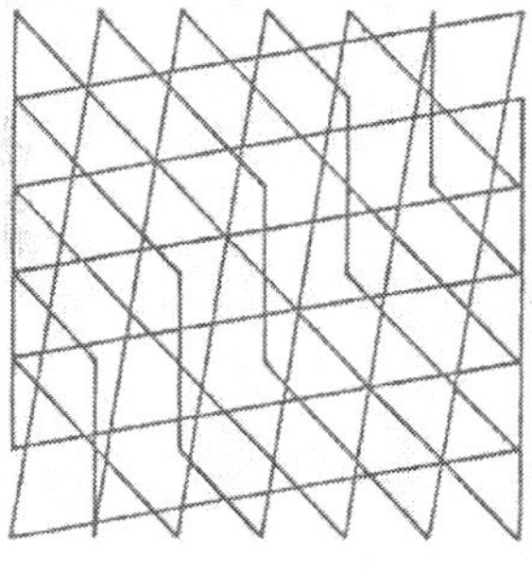

fig. 3

"Sacred Geometry is a conscious form of Ascension". (Drunvalo Melchizedek).

It is my intention to illuminate to you a most exquisite biblical parable whose interpretation can only be understood by a knowledge of spiritual mathematics. Specifically, you would need to understand what a magic square is. By definition, it is a matrix of numbers whose sum of columns, rows, and diagonals are equal. It exists as the center of the Tibetan Calendar/Cosmology.

According to ancient alphabets, like the Greek and Hebrew, every letter corresponded to a number. Thus the Magic Sum or Constant of a magic square related directly to the names of the Lord. It was for this reason that the esoteric side of magic squares was hidden from public eye.

Shown in figure 1 is the Magic Square of 7 x 7 whose sum of columns, rows, and diagonals equals 175. This mystical square came from the ancient Essenes who were a brother/sisterhood in Masada, ancient Israel who practiced fasting, celibacy and sharing of community goods in their desert lands. Of Christian interest here lies in the fact that the central crosses also add up to 175 eg. 17+49+25+1+33+7+43 and 31+7+25+43+19+49+1. When this magic square is transformed into Art, as will be shown later, the final image produced is a grand Cross (figs. 4 and 5).

Known as Sigillum XTI or the Seal of Christ, this vibrational code of Sevenness relates the planetary consciousness of Venus according to the Hebraic 'Tree of Life' cosmology that related 10 Spheres to the 10 Planets: A progression from Earth (Malkuth) No. 10 to Heaven (Kether) No. 1. This is depicted in figure 2, as the seventh Sephiroth or Sphere known as Netzach or Victory. As Venus it relates to the victory of love, the union or marriage of the lower self with the Self Divine, the tantric alchemy of raising Love out and beyond the realm of the ego and into planetary compassion for all beings regardless of creed or colour.

"This most ancient seal of XT-the High Priest, the At-one-er and At-one-ment of all things, the Lord our Lady Iesu-Maria-is one of the most ancient and sacred symbols of the Xn religion, extending back to pre-Xn antiquity, and symbolizing the restoration of the primeval order to the universe, bringing order out of confusion. In this most curious and mystical combination of numbers we have 21 summations of 175 (this number cabalistically added being symbolic of XT amid the XII), viz.,7 rows; 7 columns; 2 diagonals; 1 central cross, which sums 2 ways (5 vertical with 2 transverse numbers, or 5 transverse with 2 vertical); 3 squares made of the 4 corner numbers of the inner, middle, and outer courts with 3 vertical, or 3 transverse central numbers. There are curious combinations, eg., the sum of the 4 corner figures of each of the squares being added give 100. Likewise the extreme end numbers of the central cross, while the complimentary pairs, wherever found, being added, sum 50. Regarding the symbolism, each number denotes some place or office in the Council of the Church, whose symbol is 120, under her head or his vicar. 49 Symbolizes the number of the priesthood of the Universal Church under its head, and 25 that of the Local Church under its head, while 1 represents the Unity underlying all.

It will be noted that 49 symbolizes also the number of years in the life of XT by the testimony of His contemporaries, the Iews, and S. Irenaeus, B. M., A.D. 189, who

Page 2 of 4 cont...

The Parable of the Seven Palms

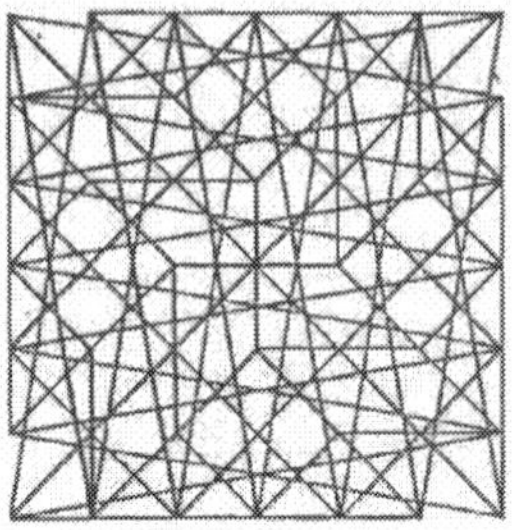

fig. 4

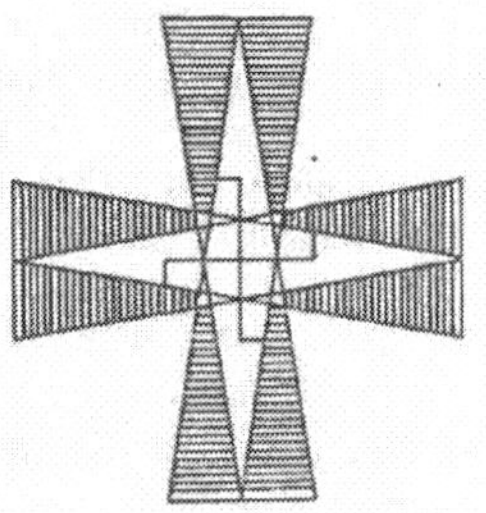

fig. 5

fig. 6

received it from the surviving apostles and disciples of the Lord. His ministry having lasted 18 years from the date of His baptism, when He was manifested as the XT, the beloved son of God." (I.H.W.Y.S.T.)

Similar to the records known as the Dead Sea Scrolls the "Parable of the Seven Palms",

fig. 1 is taken from a rare bible known as: THE GOSPEL OF THE HOLY TWELVE (in the Author's Library) also known as: THE GOSPEL OF THE PERFECT LIFE. Lection LIV p89. Verses 17-20. Translated from the Aramaic (the language that Jesus spoke) in the 18th century by 4 editors, one of them being the Master Mind Emmanuel Swedenborg. Due to times of corruption, these early manuscripts of rolled-up animal skins, found their way to foreign countries. This bible was found in a Tibetan Cave inside of a clay urn for 1700 years!! It reads the same as the King James version eg. The Lords Prayer: "Our Father-Mother who Art Above & Within... In Wisdom, Love & Equity Thy Kingdom come to All...Give us day by day to partake of thy holy Bread & the Fruit of the Living Vine... Shew upon us thy goodness that to others we may shew the same... From the Ages of Ages, NOW & to the Ages of Ages, Amun".

This exquisite parable by Jesus uses the wisdom of Majik Squares as a beautiful analogy for the DIVINE KINGDOM.

In the same bible appears a 2nd parable referring to the "Magic Square of 11 Perfected" which is also labelled: "Ecclesiae Militantis Sigillum".

How to Transform Magic Square Numbers into Art.

As simple as joining a line in sequential or consecutive order, from the first No. 1-2-3-etc to the last number 49, an energy circuit is created. This is shown as fig. 3 and technically is known as **The Essene M.Sq. of 7x7 Yantram at 0°** (Yantram is a sanskrit word literally meaning instrument or tool. Its plural is Yantra. It is a geometric, artistic power diagram that serves to act as a transistor circuit or channel conducting psychic essences).

To make the pattern more aesthetic or visually pleasing, fig. 3 can be super imposed upon itself at a 90° angle. This creates fig. 4 known as **The Essene M.Sq. of 7x7 Yantram at $0^{o} + 90^{o}$.** If you look very carefully at fig. 4, from the center and outwards, you will notice a Swaztika within a large equal-armed cross that fills the whole square. This isolation, shown as fig. 5, demonstrates how magic squares creates sacred symbols and fuses them to make powerful statements like " East & West are really One".

And finally, to demonstrate how this is a creative and experimental exercise in mathematical beauty and research, you can take the original pattern and superimpose it 8 times at regularly intervalled angles to produce the extraordinary pattern of concentric ripples, as shown in fig. 6. This is known as **The Essene M.Sq. of 7x7 Yantram at 8 x 22.5°.**

"And so the process of centering, the gathering of oneself as if by an inward throw of a stone into the pool of one's own consciousness - is also a Mandala". These words by José Arguelles are beautiful, though the word Mandala may be replaced by Yantram. The Mandala (Sanskrit=circle with center) is the invisible holographic 3 Dimensional Form, more like a sphere with a center. The Yantram the 2 Dimensional Form or circle Art that is visible, upon which when meditated upon, assists the viewer to grasp the invisible mandala or Formlessness.

As you meditate upon fig. 6, treat it as you would a crystal and program it with specific thoughts of Love, Truth, Harmony, Beauty, and Balance; qualities that you desire to crystallize into your life, now.
Remember that in the history of Magic Squares, like in the medieval times, some folk tended to corrupt the highest unselfish aspects of these harmonic codes and used them to destroy enemies (eg. Mars). This, by the cosmic Law of Boomerang, would also bring mishap to the user, eg. if a man wanted Mary to fall madly in love with him, even against her will, he would select the Venusian Magic Square of 7x7, as a Talisman of Love, and use the letters of her name: Letting M=4, A=1, R=9, Y=7 (according to the simple Pythagorean numerological system where A=1, B=2, C=3 etc.). He would thus draw an unbroken line from 4 to 1 to 9 to 7 and fill the circuit with his thoughts as if to amplify his desire. This is the petty use of such mystical numbers.

May this article lead you into a deeper research of all The Lost Knowledges. It appears to me that all the Lost Metaphysical Sciences, especially those that were protected and hidden from the masses eg. Reiki, Merkabah Geometries, Magic Squareology, Masonic Teachings, Free Energy Systems (Tesla), Colour and Sound Therapies etc.. the list is endless, all these spiritual sciences are ready to be shared to assist us all in these accelerated times, to handle with ease the imminent physical Ascension of:
All That You Are
The God I Am
The Unlimited God I Am.

Form Is The Envelope of Pulsation
[Ancient Tantric Aphorism J.

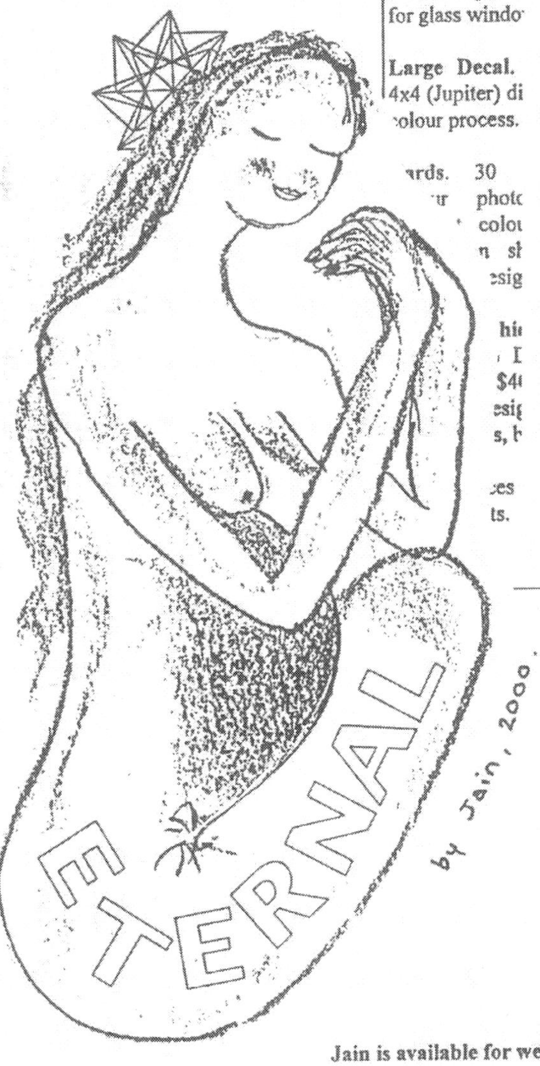

Article submitted to 'Sedona' magazine of New Zealand April 1997 Sydney.

Jain is available for weekend seminars.

ALL CORRESPONDENCE TO: 6 1 8
7 5 3 2 9 4

THE BOOK OF MAGIC SQUARES VOL 3 YEAR 2000 JAIN

- This unique and original "Draw-Your-Own-Mandala-Join-The-Dots-Colouring-In-Activity-Book is equally suitable for children and adults. Just by drawing a long, continuous line from 1→2→3 etc to the last No. of a Magic Sq., superb patterns appear, creating sacred symbols from many cultures and verging on Atomic Art.
- The traditional meaning of a M.Sq. is that the sums of its columns, rows and diagonals are Equal.
- Volume 1 (pub. 1990) covered M. Sqs. of 3, 4, 5, 6 and 7.
- Volume 2 (pub. 1996) covered M. Sqs. of 8, 9 and 10.
- Volume 3 (pub. 2000) covers M. Sqs. of 11, 12 and 16, in detail, giving many varied and creative ArtForms depicting Order amongst Chaos.

- At the end of this Volume 3 there is a chapter giving all the graphical Solutions, as do Volumes 1 and 2, but unlike the others, this Volume has a special and in-depth Glossary of all Sacred Geometrical terms used.
- Also, unique to this Volume, is a chapter on "The Visual Multiplication Table and its Reduction Grid" (also known as The Vedic Square), that uses digital sums to create Atomic ArtForms, like the nuclear geometry of Platinum Crystal (shown here) for the first time in print, linking 20 years of Magic Square research to the lost Vedic Mathematical Sutra that Jain calls "... in the Next Dimension".
- Explore, extract what you can from this quarry of information and have Fun.

Jain Products Order Form

NEW! **The Living Mathematics of Nature, Series – 5 DVD Set, pub: 2006**

	Item	Price	Weight
☐	**Complete 5 DVD Set** (Special! Save $25)	**$200**	240 gm
☐	**Introduction to Ancient Knowledge** (DVD #1 of 5)	**$ 45**	100 gm
☐	**Vedic Mathematics** (DVD #2 of 5)	**$ 45**	100 gm
☐	**Magic Squares** (DVD #3 of 5)	**$ 45**	100 gm
☐	**The Divine Phi Proportion** (DVD #4 of 5)	**$ 45**	100 gm
☐	**3-Dimensional Geometry** (DVD #5 0f 5)	**$ 45**	100 gm

Magic Squares

	Item	Price	Weight
☐	**The Book of Magic Squares, Volume 1** (M. Sqs. of 3, 4, 5, 6, 7)	**$ 45**	520 gm
☐	**The Book of Magic Squares, Volume 2** (M. Sqs. of 8, 9, 10)	**$ 45**	520 gm
☐	**The Book of Magic Squares, Volume 3** (M. Sqs. of 11, 12, 16)	**$ 45**	560 gm
☐	**Join-The-Dots** (adults and children coloring book. Pub:1999, 56 pp)	**$ 30**	160 gm
☐	**Decal: The Magic Square of 4 X 4 (Jupiter)** (Small, Magenta)	**$ 10**	20 gm
☐	**Decal: Magic Square 8 X 8 (Mercury)** (Small, Blue Background)	**$ 10**	20 gm
☐	**Decal: Magic Square 4 X 4 (Jupiter)** (Large, Yellow Background)	**$ 20**	100 gm
☐	**The Gospel of the Holy Twelve** (Essene Bible w. Magic Sqs. 210pp)	**$ 45**	410 gm

Vedic Mathematics

	Item	Price	Weight
☐	**The Magic of Nine in Vedic Mathematics.** (Pub:1999, 124pp)	**$ 45**	390 gm
☐	**The Vedic Mathematics Curriculum for the Global School. Part 1, Digital Sums.** (Pub:2001, 142 pp)	**$ 45**	460 gm
☐	**Jain Mathemagics Curriculum for the Global School. Part 2. Multiplication.** (pub: 2005, 152pp)	**$ 45**	480 gm
☐	**Vedic Mathematics for the New Millennium: DVD** (Pub:2002)	**$ 45**	100 gm
☐	**Dolphin Human Connections:** Jain shows Vedic Maths (2006). 40 mins	**$ 50**	100 gm
☐	**Vedic Mathematics** by: Bharati Krsna Tirthaji **Soft Cov.** (355pp)	**$ 45**	460 gm
☐	**Vedic Mathematics** by: ibid. **Hard Cover.** (Pub:1965, 355pp)	**$ 55**	570 gm
☐	**Calculations @ The Speed Of Light** by Kranti Kiran (pub:2002, 104pp)	**$ 45**	230 gm

Sacred Geometry

	Item	Price	Weight
☐	**The Book of Phi, Vol 1** (The Living Maths of Nature. Pub:2002, 172pp)	**$ 50**	550 gm
☐	**In The Next Dimension** (The Book of Phi, Vol 2. Pub:2003, 188pp)	**$ 50**	580 gm
☐	**Propositions on the Exact Value of Pi.** Reddivari Sarva J. Reddy	**$ 60**	620 gm
☐	**Peter's Projection Map.** (Earth's true cartographic proportions)	**$ 25**	50 gm
☐	**Dvd: Jonathan Quintin's "Visual Symphonies"** Toroidal Animation	**$ 45**	100 gm

EarthHeart Jewellery

	Item	Price	Weight
☐	**Platonic Solids:** Crystal Gift Set in wooden camphor boxes	**$100**	50 gm
☐	**Heart Star:** 24 carat gold plated stellated cuboctahedron	**$150**	50 gm
☐	**Tantric Star:** 24 carat gold plated Star tetrahedron	**$130**	50 gm
☐	**Solar Star:** 24 carat gold plated stellated dodecahedron	**$150**	50 gm

SUB-TOTAL: $ ....

Postage & Handling: $ ....

TOTAL: $ ....

Tax Invoice

Name: ....

Mailing address: ....

City/Town: .... **State:** .... **Zip Code:** ....

E-mail Address: ....

Home phone: .... **Mobile:** ....

Please choose your preferred method of payment:

☐ **Cash, Cheque or Money Order.** Amount: ....
Payable: JAIN F.R.E.E.D.O.M.S.

☐ **PayPal now Available!** Amount: ....
Email: jain@jainmathemagics.com

☐ **Direct Deposit** Amount: ....
Acc Name: JAIN F.R.E.E.D.O.M.S. BSB: 032573 Acc Number: 171268

☐ **(circle) Visa / Mastercard** Amount: ....

Card number..

Name on Card: .... **Exp. date:** ....

AUSTRALIAN POSTAGE: Orders up to $50, add $8. Orders $50 to $120, add $12.

INTERNATIONAL ORDERS: $10 set up fee and add $10 per item. (Approx.)
International Bank Draft in Australian Dollars. PayPal also available + Western Union.
Cheques made out to JAIN F.R.E.E.D.O.M.S.

All Prices are listed in Australian Dollars. **All products available online!**

Date Of Order .... **Dispatched** ....

6	1	8
7	5	3
2	9	4

JAIN F.R.E.E.D.O.M.S. inc 9882763 trading as JAIN MATHEMAGICS
For Research Expressing Essential Data Of Magic Squares
Phone: +61 02 6684 4409 Mobile: +61 0423 583 886 ABN 47 244 872 696
Email: jain@jainmathemagics.com Web: www.jainmathemagics.com
Address: 777 Left Bank Rd, Mullumbimby Creek, NSW, 2482, Australia

JAIN MATHEMAGICS CURRICULUM
for the GLOBAL SCHOOL

formerly known as: The VEDIC MATHEMATICS CURRICULUM for the GLOBAL SCHOOL

Book 2 MULTIPLICATION JAIN 2005

Developing the Inner Mental Screen

JAIN MATHEMAGICS CURRICULUM FOR THE GLOBAL SCHOOL Book 2 MULTIPLICATION 2005

"THE FUTURE IN MATHEMATICS IS DOING IT IN YOUR HEAD" **(by Jain: *Mathematical Futurist*).**

- This is an Educational Workbook for Children and a Teacher's Resource Material for those interested in learning and teaching Rapid Mental Calculation. It increases the child's Memory Power and Confidence.
- This long-waited for book of Ancient Mathematical Short-Cuts excavates many hidden truths and vital properties in the playing field of Numbers.
- There is a rare and unique chapter called "HARMONIC STAIRWAY" from Jain's Dictionary of Numbers, for the first time in print. These choice Mathematical Plums are something that the student stumbles upon. They are working away solving calculations mentally, and when they check their answers in Chapter 2, they are instructed, in the section: DID YOU KNOW, to learn more about that particular number. But these are not any particular numbers, they are Anointed Numbers that our forebears held in high esteem.
- Beautifully and richly illustrated that the graphics alone educate. The student learns about other Families of Numbers, other numerical relatives and mathematical cousins that enrich their understanding of what mathematics is really about. It instills that sense of joy and wonderment that Pythagoras and Baudahayana knew.
- There is no error; it is an infallible bulletproof system, based on Unity Consciousness. The 16 "Threads" or Sutras that solve all known mathematical problems express The Law of Economy and The Path Of Least Resistance.

- As Jain Mathemagics becomes more globally acknowledged, it will help end the generational tyranny that has kept such knowledge in the dusty cupboard. This book is an organic pill that will prevent the slowly encroaching borgificiation of Mathematics.
- Since the turn of the bi-Millennium, there has been a global renaissance in the subject of "Sacred Geometry". You could summarize it as a fascination for the Language of Shape. This book invites seekers to truly incorporate SHAPE into their ability to perform mathematics, to apply this Language of Shape to the next octave of learning, not theoretically, but by practical use of using SHAPE to literally perform mental calculations in seconds.
- Often students come out of school unable to recover from deep mathematical wounds. They are mis-diagnosed as "un-intelligent" or "dumb", given drugs like Ritalin, told they are ADHD (which really means: Attuned Directly to Higher Dimensions). In fact, Dyslexic children are geniuses.
- **UNCOVER, RECOVER, DISCOVER**

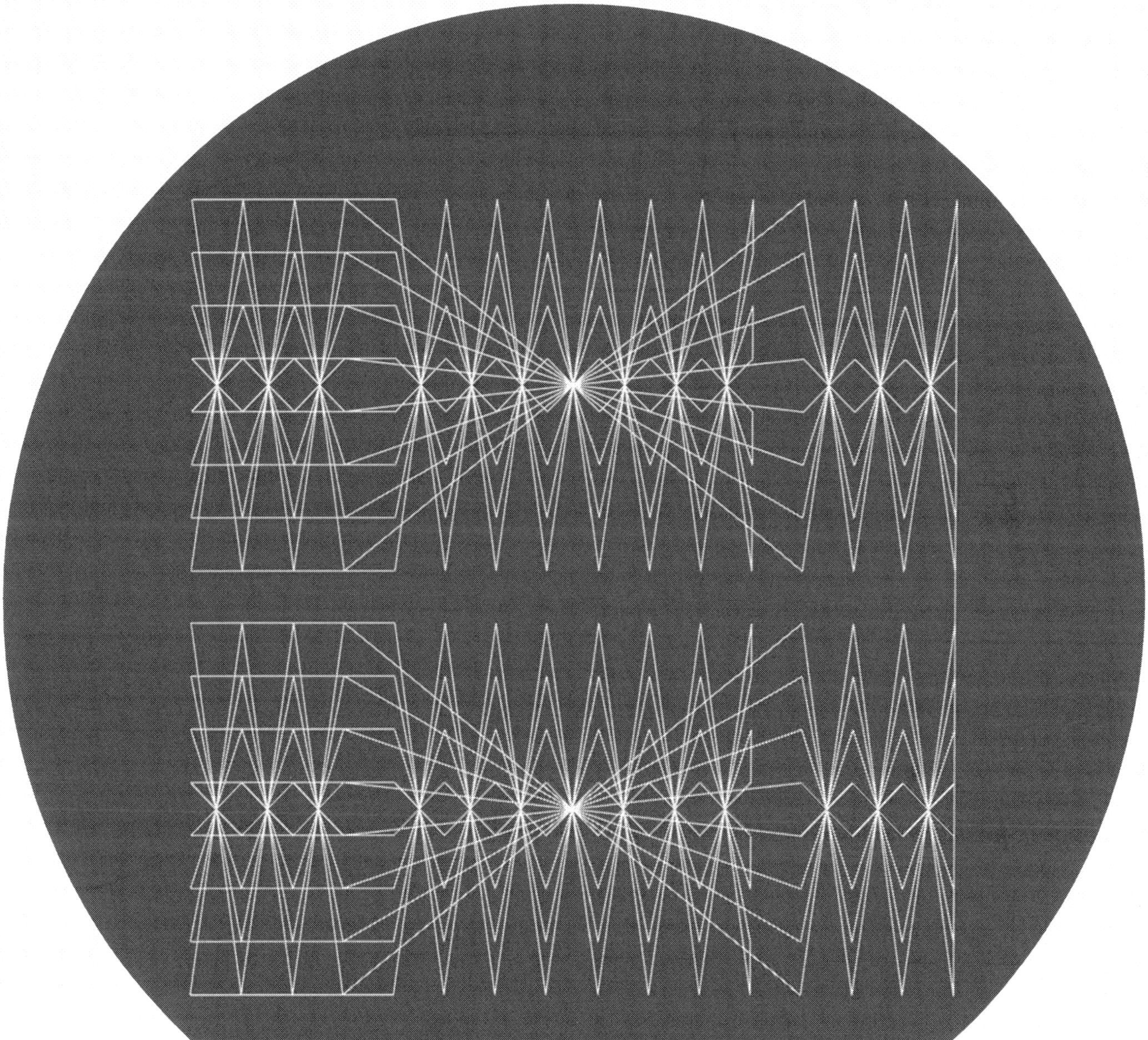

Made in United States
Orlando, FL
04 November 2026

90200387R00103